TH　　　　　　　　　　　　　　　　　PAIR

F

AN

C

H　　　　　　　　　　　　　　　　EDITION

THE FIBERGLASS REPAIR
AND CONSTRUCTION HANDBOOK

2ND EDITION

JACK WILEY

TAB Books

Division of McGraw-Hill

New York San Francisco Washington, D.C. Auckland Bogotá
Caracas Lisbon London Madrid Mexico City Milan
Montreal New Delhi San Juan Singapore
Sydney Tokyo Toronto

pbk   10 11 12 13 14 15 DOC/DOC 0 9 8 7 6 5 4 3 2
hc   2 3 4 5   6   7   8   9 10 11 DOH/DOH 9 9 8 7 6 5 4 3 2 1 0

Library of Congress Cataloging-in-Publication Data

Wiley, Jack.
    The fiberglass repair and construction handbook.

    Includes index.
    1. Glass reinforced plastics. 2. Glass fibers.
    I. Title.
    TA455.P55W5 1988   666'.157   88-2162
    ISBN 0-8306-0279-8
    ISBN 0-8306-2779-0 (pbk.)

Front cover photographs (clockwise) from top left) courtesy of: Kohler Company Inc., Kohler WI: Century Boat Company, Manistee, MI; and Chervolet Motor Division, General Motors Corporation, Warren, MI

                                                            HT3
                                                            2779

Contents

Mat, Cloth, and Woven Roving Laminates Bonded to Wood—Mechanically Fastening and Bonding Open Weave Fiberglass Cloth to Wood—Mechanically Fastening and Bonding Polyester Cloth to Wood—Mat and Polyester Resin Laminate with a Core—Cloth and Polyester Resin Laminate with a Core—Mat, Cloth, and Polyester Resin Laminate with a Core—Comparison of Core Laminates—Bonding Laminates to Various Materials with Polyester Resin—Polyester Putty and Fillers—Sanding Polyester Laminates—Summary of Fundamental Techniques of Fiberglassing with Polyester Resin—Facts on Epoxy Resins—Types of Epoxy Resins—Adding Curing Agent or Hardener—Adding Reinforcing Material to Epoxy Resin—Adding Epoxy Resin to Mat—Adding Epoxy Resin to Cloth—Laminating Two Layers of Mat with Epoxy Resin—Laminating Two Layers of Cloth with Epoxy Resin—Other Methods of Laminating with Epoxy Resin—Using Epoxy and Polyester Resins in the Same Laminate—Other Epoxy Resin Laminates—Adding a Mat and Epoxy Resin Laminate to Wood—Adding a Cloth and Epoxy Resin Laminate to Wood—Adding Polyester Layers to Epoxy Laminates—Other Methods for Adding Epoxy Resin Laminates to Wood—Bonding Laminates to Various Materials with Epoxy Resin—Epoxy Putty and Filler Compounds—Sanding Epoxy Laminates—Summary of Epoxy Practice Exercises—Basic Contact Molding with Polyester Resin

Introduction

Fiberglass is a combination of glass fiber reinforcing material and plastic resin. It is used in the construction of boats, car bodies, campers, travel trailers, motor homes, tanks, swimming pools, hot tubs, sinks, bathtubs, surfboards, outboard motor covers, furniture, toys, snowmobile bodies, shower stalls, sporting equipment, and hundreds of other things.

In spite of some claims to the contrary, fiberglass often suffers cosmetic and structural damage. This book fills the need for how-to information on repairing this damage. Materials, tools, safety, and techniques are covered in detail for repairing many types of damage to things made of fiberglass.

Complete information is given for using fiberglassing to repair and protect items made from other materials, such as wood and metal. There's a large section on using fiberglassing for metal automobile body repair and customizing work. The possibilities for assembling fiberglass boats, car bodies, and other things from manufactured kits are also explored. Fiberglass construction techniques and projects are introduced.

Once you master the basic skills and techniques, fiberglassing repair work is quite easy. This book will help the do-it-yourselfer do his or her own fiberglassing repair work to save money (having fiberglassing done by a professional is expensive). It can also serve as a basis for anyone who wants to set up a fiberglassing repair business. It is suitable as a basic text for classes and courses in fiberglassing. This book also gives a solid foundation in fiberglassing skills for those who want to go on to more ambitious projects, such as the actual construction of fiberglass boats, car bodies, fish ponds, and even original creations.

Fiberglassing is considerably different from making things out of wood, metal, and other existing materials. Fiberglassing has a chemical aspect. You take a liquid resin, add a catalyst or curing agent, combine this with glass fiber reinforcing material, and come up with hard fiberglass reinforced plastic material. It seems like magic when you first see it happen.

While much of the material in this book comes from my own fiberglassing repair projects and experiences, many other people shared their ideas and techniques with me. Thus, it is impossible to acknowledge all the help given to me. To the many people who freely shared their ideas, I want to extend a sincere thanks.

Chapter 1
Fiberglass Facts

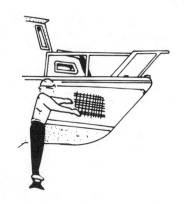

The term *fiberglass* is somewhat of a misnomer. It originally referred to thin glass filaments. While fiberglass is still used to refer to these glass fibers, as well as feltlike mat and woven fabrics made from these fibers, it is now also commonly used to refer to a two-part structural unit consisting of glass fibers combined with a resin (usually polyester and sometimes epoxy).

In order to prevent confusion, the term *fiberglass reinforced plastic* (FRP) is sometimes used to refer to the two-part structural unit, with the "Fiberglass Reinforced" referring to the glass fiber reinforcement and the "Plastic" to the resin. This terminology is generally not used outside the plastics industry. Most people refer to a boat, hot tub, swimming pool, car body, and many other things made out of fiberglass reinforced plastic as being fiberglass.

By the end of World War II, the chemistry and materials for fiberglass had developed to the point where practical applications were possible. The United States Navy and United States Coast Guard used fiberglass construction for personnel boats. The use of fiberglass soon spread to the private sector, not only as a boatbuilding material, but also for furniture, lampshades, tanks, tubs, and even car bodies.

Over the years, fiberglass has been tried as a substitute material for wood, metal, concrete, and other materials. When the first small fiberglass pleasure boats appeared on the market after World War II, almost all small pleasure craft were made of wood. The vast majority of pleasure boats used in the United States today are constructed from fiberglass.

As a construction material, fiberglass has both advantages and disadvantages. For some uses it proved feasible; for others it didn't. In some cases it was too costly. For some uses other materials had superior structural qualities.

Fiberglass is often combined with other materials to take advantage of desirable characteristics of fiberglass while avoiding its undesirable aspects for the particular application. For example, fiberglass car bodies might use metal fasteners, hinges, latches, and trim and reinforcing members. Fiberglass boats usually consist of hulls, decks, and cabin structures of molded fiberglass, often with wood or foam pieces molded in. Metal fasteners are commonly used along with chemical bonding to join the various moldings together. Many components of wood, metal, and various plastics besides fiberglass are used for completing the boats. Fiberglass is generally unsuitable for engine and mechanical construction.

Thus, finished fiberglass laminates or moldings (fiberglass reinforced plastic) are a matrix of glass fibers and resin. These are, in turn, combined with other materials to make workable and finished things or decorative items.

The widespread use of fiberglass has created a demand for fiberglass repair work. In spite of some claims to the contrary, fiberglass will deteriorate if not properly maintained, and it can be damaged by a variety of means. Fiberglass repair businesses are now found in most parts of the United States.

1

Even so, there is an extreme shortage of qualified fiberglass repair workers in most parts of the country. Those people who are in business can demand and get large amounts of money for repair work. This is good if you are being paid to do the repair work, but not so good if you are paying someone to do it for you. Compared to most metal and woodwork, relatively few tools and little equipment is required for typical fiberglass repair work.

THE RELUCTANCE TO DO FIBERGLASS REPAIR WORK

There has been a reluctance on the part of many people to attempt their own fiberglass repair work, even though they might routinely make repairs involving other materials around their homes or on their automobiles. One reason is the chemical aspect of fiberglassing. This is quite different from taking a piece of wood and cutting and shaping it to replace a piece of wood that has been broken or damaged. Fiberglassing involves mixing chemicals together that are in liquid form, applying them to fiberglass reinforcing material, and by means of chemical reactions that take place within the liquid, coming up with a hard fiberglass laminate. While the exact chemistry is a mystery even to most experienced fiberglassers, the fact that a hard fiberglass laminate results is almost unimaginable to those who have not seen or done any fiberglassing.

A second reason why more people don't attempt their own fiberglass repair work is the lack of how-to information. What tools, materials, and supplies are needed? How does one go about learning the techniques? How are various types of damage best repaired? This book is an attempt to provide the needed information.

Health and aesthetic considerations are also responsible for the reluctance of many people to attempt fiberglass repair work. Precaution must be taken to guard your health and safety when fiberglassing. The chemicals can be sticky and messy, and fiberglass reinforcing materials make your skin itch. With proper handling and use of protective clothing and equipment, it's possible to avoid these problems, too.

THE PROFESSIONAL AND THE DO-IT-YOURSELFER

Once you learn the basic techniques and safety procedures, fiberglassing can be enjoyable, challenging, and creative. The techniques presented in this book are intended primarily for the do-it-yourselfer, although they can easily be adapted for a fiberglassing repair business. There are two important differences. The do-it-yourselfer usually doesn't have to consider time as money, whereas the person doing it to earn money does. In most cases the professional has to work rapidly and efficiently to maximize the profits from his or her labor. With some experience and practice, the do-it-yourselfer can often achieve the same professional results. It will almost certainly take much longer to do the job.

A second important difference is that the professional has more and better tools and equipment than the typical do-it-yourselfer. The emphasis in this book is on doing quality work with a limited amount of tools and equipment because the do-it-yourselfer usually has to absorb the cost of the tools and equipment with a limited number of repair jobs. The professional can spread this cost over hundreds of jobs.

REPAIRING DAMAGE TO FIBERGLASS

Many repairs can be made to fiberglass laminates that have been damaged from accident, wear, abuse, lack of maintenance, or deterioration. An important part of fiberglassing is to evaluate the extent of the damage and know if the repair is within your skill level. Several practice exercises are given so that the basic skills and techniques of fiberglassing can be learned before an actual repair is attempted.

In most cases repair to fiberglass will be made by fiberglassing. For example, if there is a hole in the fiberglass that needs to be filled in, the repair will be made by fiberglassing—that is, by adding catalyst to liquid resin and applying this to fiberglass reinforcing material. If done properly, this bonds to the original fiberglass and forms a hard fiberglass laminate to fill in the hole.

In some cases, however, other methods and materials will be used for making the repairs, depending on the particular damage and the way the fiberglass is used. Sometimes more expensive reinforcing materials, such as polypropylene, polyester, and acrylic fabric, will be substituted for fiberglass reinforcing material. Frequently, polyester and epoxy repair compounds that do not contain any glass fibers will be used. Some repairs can best be made by mechanical means such as bolting a piece of fiberglass, wood, metal, or other material in place to cover the damaged area of fiberglass. An important fiberglassing skill is the ability to determine the most practical approach for repairing a particular damage.

UNDERSTANDING HOW THINGS ARE MANUFACTURED FROM FIBERGLASS

A basic understanding of the molding and lay-up methods used for manufacturing fiberglass products is useful for repairing these things later when they are damaged. These methods are detailed in later chapters. After the component fiberglass parts have been molded, various methods are used for assembling them into finished products. This involves bonding and/or mechanical fastening.

BUILDING WITH FIBERGLASS PANELS

Both flat and corrugated fiberglass panels are now available. These can be used for building patio roofs, as siding, for making greenhouses, and for many home improvements. Accessories such as aluminum nails with neoprene washers, caulking, sealer, wood strips with precut corrugations, and flashings are available. Various skills and techniques for working with cured fiberglass, such as sawing, drilling, and filing it, are detailed in later chapters. Combining fiberglassing methods will greatly increase the possibilities for using fiberglass panels.

PROTECTING AND REPAIRING OTHER MATERIALS WITH FIBERGLASS

A second important repair aspect of fiberglassing is to use it to protect and repair damage on things made out of other materials. For example, plywood boats can be sheathed on the outside with fiberglass for protection and added structural strength, or even to repair damage to the wood. Metal gutters can be repaired by fiberglass reinforcing material and epoxy resin. Many plastics can be repaired by fiberglassing techniques.

While damage to many materials is best repaired by more conventional means, there are many situations where fiberglassing is the way to go. Because bonding is frequently a problem when fiberglassing to wood, metal, and many types of plastics, epoxy resin—which has better bonding qualities—is often used instead of the less expensive polyester resin. Preparation of the bonding surface is also extremely important.

The techniques for protecting and repairing wood, metals, and many plastics are detailed in later chapters. The application of fiberglassing techniques for making repairs to metal auto bodies is extremely useful. When properly done, sound and lasting repairs are possible without the equipment and tools required for conventional auto body work.

FIBERGLASS COMPONENTS AND KITS

Many fiberglass components and kits are now on the market, such as replacement fiberglass car fenders and hoods. Most of these components are installed by bolts and other mechanical fasteners; no fiberglassing is required. There are also fiberglass customizing items, such as air scoops and fender flairs, that are readily available. Some of these are bolted in place; others require fiberglassing techniques for installation.

Fiberglass hot rod and sports car bodies are also available. These are usually assembled without fiberglassing, although some fiberglassing might be necessary in some cases.

Many fiberglass boats are available in kit form. In most cases all of the molded fiberglass components are supplied, so the builder does not have to do any fiberglass molding. Fiberglassing skills are generally necessary for assembling the components. For example, fiberglass bonding strips are often used in addition to mechanical fasteners to join fiberglass moldings, such as the hull to the cabin and deck molding, together as a single unit. Plywood bulkheads and other structural members are commonly attached to the fiberglass moldings with fiberglass bonding strips and angles. Thus, a solid foundation in fiberglassing skills will help to assure proper assembly of these kits.

CUSTOMIZING AND MODIFYING

In addition to repair work, fiberglassing skills and techniques can be used to customize and modify work. For example, car bodies and boats can be customized and modified in many creative ways. In a sense, fiberglassing materials are used as the clay to carry out your ideas. You might want to make something look better, strengthen it, or make it work better. Fiberglassing can often be a method for accomplishing these things. Fiberglassing can be used for customizing and modifying not only things made out of fiberglass but also those items constructed from wood, metal, and many other materials.

BASIC FIBERGLASS MOLDING AND CONSTRUCTION TECHNIQUES

Many people who learn basic fiberglassing repair work will want to go on to fiberglass molding and construction techniques. The fundamentals are covered in later chapters of this book. These techniques can be applied not only to more advanced repair work but also to make things "from the ground up." You can make a mold from an existing air scoop, car fender, or other item and then make a duplicate from fiberglass. An old wood dinghy can be used as a plug for making a mold for a fiberglass dinghy. You can then use the mold for laying up or making one or more (the mold can be used over and over again) dinghies out of fiberglass.

There are also several "one-off" construction methods that do not require a female mold. These can be extremely useful when only one of an item is required. Many methods for one-off fiberglass construction are covered in later chapters. These are useful for repair work and for constructing things.

GEL COATS AND PAINTING

There are two main aspects of most repair work: structural and cosmetic. The cosmetic part is the more difficult to do well. Take the case of a break in the gel coat surface of a fiberglass molding. Fiberglassing in the damaged area to the point where it is as strong as or stronger than the original molding is fairly easy. Sanding and refinishing the gel coat surface to conceal the fact that a repair was made can be quite difficult. The same thing applies to a fiberglass repair on a metal automobile body. Touch-up painting to match that of the rest of the car body can be difficult.

For this reason, considerable emphasis is placed on sanding, surfacing, and finishing fiberglass surfaces with gel coats and various types of paints. I've seen many do-it-yourselfers spoil otherwise satisfactory fiberglassing repair work in the refinishing stages. Sometimes the surface isn't smooth and fair before the gel coat or paint is applied. In other cases the touch-up gel coat or paint does not match the old finish. Perhaps the do-it-yourselfer did not take the time and care necessary to do the job right, or did not have enough skill and experience, or did not use the proper materials.

Many beginners will spend considerable time with scrap materials learning to lay up a fiberglass laminate, but they will attempt refinishing work on an actual repair job with little or no previous practice on scrap materials. To achieve professional results in making fiberglassing repairs, finishing techniques must be taken as seriously as fiberglass lay-up work.

REASONS FOR DOING YOUR OWN FIBERGLASSING REPAIR WORK

Before going on to actual fiberglassing repair work, I think it is important to take a look at some of the reasons for doing this work yourself.

Save Money

Perhaps the most important reason for doing your own fiberglassing repair work is to save money. Having fiberglassing work done for you is expensive. By doing your own work, you save money not only by providing your own labor but by purchasing materials and supplies from discount supply firms yourself. Most fiberglass repair workers charge $10 or more an hour. To add to their profits, they buy the materials wholesale or at dealer discount prices and charge the customer retail prices.

Get the Work Done the Way You Want It

By doing your own repair work, you have the opportunity to do the job the way you want it. The professional fiberglasser is likely to look at the job you need done only in terms of how much profit can be made for the job. Once an estimate is given, the worker is likely to attempt to do the job in the shortest amount of time possible, with the minimum expense for materials used. This often means cutting corners. If the repair job is a small one, it's often difficult to find a professional who wants to bother with it.

Also, it's important to keep in mind that, while some fiberglassers have had years of experience, others are just starting out. Because there are no set qualifications or standards for a fiberglasser to go into business, it is often difficult to judge his or her competency.

By doing your own repair work, you can get around these problems. You can take the time to do the job the way you want it done, provided you are willing to first take the time to learn the necessary skills and techniques.

Get the Work Done When You Want It Done

Getting repair work done when you want it done can be difficult, especially if it is a small job. This applies to TV, auto, home, and practically every other type of repair. It seems especially true of fiberglassing repair because there is a shortage of workers in most parts of the United States.

Avoid Getting Ripped Off

While the rip-offs and rackets in fiberglassing repair work are less widespread than in auto repair, they do exist. Most people have little idea how much a particular fiberglassing repair should

cost, so they are open to price gouging. You can avoid this by doing your own repair work, but in exchange, you put your own skills and talents on the line. By taking the time and trouble to learn the basic skills and techniques, difficulties should not be encountered.

The Challenge and Satisfaction of Doing It Yourself

Some people only do repair work to save money, but others relish the challenge and satisfaction from doing this work themselves. If repairing a car fender, you might want your finished work to look as though no damage had ever been present or repair made. If assembling a fiberglass car body or boat kit, you can show off the results of your efforts.

Some people will want to go on to more advanced fiberglassing and make things from fiberglass. There are many possibilities that can result in considerable challenge and satisfaction from doing it yourself.

Hobby and Craft Possibilities

Fiberglassing can be a meaningful hobby, and many hobby and craft items can be fabricated from this material. You can, for example, create original lampshades and mobiles.

FIBERGLASS AS A CONSTRUCTION MATERIAL

Before going on to actual fiberglassing techniques, a look at the advantages and disadvantages of fiberglass as a construction material is in order. For some uses, fiberglass has proven itself to be a superior or at least competitive material as compared to other possible building materials. For other uses, fiberglass could not compete — for one reason or another.

Advantages

Cured fiberglass (fiberglass reinforced plastic) does not corrode or rot. It is inert to most common chemicals, including most fuels and common pollutants. This makes it ideal as a boatbuilding material and for other uses where it will be subject to severe environmental conditions. A steel car body, for example, will rust; one of fiberglass won't. There are several other factors involved that have kept fiberglass from being more widely used as a material for manufactured car bodies. The fiberglass bodies on Chevrolet Corvettes show that the material is practical at least for luxury sports cars.

The fact that fiberglass is chemically inert to most common chemicals means that it is not susceptible to *electrolysis*. This is an important advantage for use in or around salt water or marine environments.

Fiberglass is strong and lightweight. It has approximately the same structural strength as wood or steel weighing twice as much when properly designed and fabricated from quality materials.

Fiberglass can be molded in one-piece seamless forms that are leakproof, making it ideal for boat hulls and deck and cabin structures, roofs for a variety of recreational vehicles, swimming pools, bathtubs, hot tubs, shower stalls, tanks, and other things. There are, however, some limits to the shapes that can be molded using standard molding techniques. Many things constructed from fiberglass do have seams and lap joints. For example, most boats have separate hull and deck/cabin structures that are then bonded and mechanically fastened together.

Fiberglass can, however, be molded into complex shapes. This is fairly easy to do and allows shapes and designs that are difficult to achieve from wood and metal.

By means of a resin gel coat, colors can be molded in. The color might fade in time and require cosmetic painting.

Fiberglass is durable. When properly fabricated, fiberglass laminates will last for 30 years, and perhaps much longer, with no significant degradation in laminate properties. Some fiberglass boat hulls constructed 30 years ago are still in sound condition, and they can have many more years of useful life. Fiberglass requires relatively little maintenance because of its noncorrosive nature.

Compared with most materials that can be used for similar applications, fiberglass is relatively easy to repair. This fact is not well known. Even major damage can be repaired fairly easily to the point where it is as strong as, or stronger, than it was originally. Fiberglass boats and other things, even though badly neglected for many years, can often be restored. You don't have to worry about rot, as you do with wood, or rust, as you do with steel.

Fiberglass can be engineered to make maximum use of the material. Various laminate thicknesses can be used on the same molding to give strength and stiffness where it is needed. A thinner laminate can be used to save weight in areas where less strength and stiffness is adequate.

There is little waste of materials in most fiberglass molding methods. Most of the resin and reinforcing material ends up as part of what is being made. In wood construction, for example, there are usually scrap pieces that do not end up in the finished product. The same applies to many types of steel construction. In the case of wood and steel, standard cuts and sizes are usually used. These must be cut and shaped for the particular application. Fiberglass molding is largely a chemical construction method, rather than building from preformed materials.

Disadvantages

Fiberglass laminates have less stiffness than steel and even aluminum. This can cause problems in applications were deflections can cause failure. Various stiffening members can often be added to the fiberglass laminate to get around this problem.

The fatigue strength of fiberglass is lower than that of steel, which can cause problems in areas where stress concentrations are located. The buckling strength of fiberglass is also low.

Fiberglass has less abrasion resistance than steel. The abrasion resistance is better than that of wood, however. In fiberglass areas where chafing can occur, various types of protective devices made from metal, rubber, or other materials can be used to get around this problem.

Unless special, more expensive, fire-retardant resins are used, fiberglass is vulnerable to fire. It is similar to plywood in that respect.

The high cost of fiberglass construction is

generally considered to be a disadvantage of this material. The cost of steel and wood construction is less on a pound per pound basis than fiberglass. For many applications, though, various desirable features of fiberglass make it worth the additional cost.

Construction of molds and tooling for fiberglass manufacturing can be costly. This cost can be amortized if the same mold is used for a large number of moldings. Some molds, for example, have been used to make 1,000 or more boat hulls from a single mold. Various one-off fiberglass construction methods have now been developed that do not require expensive molds and tooling. These are generally used for custom construction rather than for regular manufacturing.

Chapter 2

Materials

Several materials are used for fiberglassing repair and construction work. Many of these are used for manufacturing boats, swimming pools, car bodies, and other things from fiberglass (see Chapter 8).

REINFORCING MATERIALS

Although reinforcing material made from glass fibers is the most commonly used, it's not the only one. Many new reinforcing materials have been developed. For certain uses, these have qualities superior to glass fibers. Even though a laminate might have no glass fibers in it, it is still commonly referred to as fiberglass. I will follow this practice in this book. The reinforcing material is one part of a two-part structural unit that forms fiberglass (fiberglass reinforced plastic).

Glass Fiber Reinforcing Materials

Glass fiber is a fairly new synthetic dating back to the early 1930s. Lime, alumina (aluminum oxide), and borosilicate are used to form "E" glass. The "E" glass is melted and formed into continuous filaments by mechanical attenuation. Molten glass is pulled through small holes in the bottom of an electrical furnace and stretched into thin fibers, which are made into yarns. These yarns are then formed into glass fiber reinforcing materials by weaving on regular weaving machinery and other means.

To make the glass fiber reinforcing material suitable for use with plastic resins, special finishes such as chroming are added to the glass fibers. These allow the plastic resins to flow around the glass fibers and minimize trapped air. Glass fiber reinforcing material is available commercially in many forms, including cloth, mat, woven roving, chopped strands, combination chopped strands and woven roving, and milled fibers.

Glass Fiber Cloth

Glass fiber cloth is shown in Fig. 2-1. Threads made of glass fibers are woven into cloth using regular textile weaving machinery. Many different weaves are used, with the plain weave (Fig. 2-2), long shaft satin weave (Fig. 2-3), and unidirectional weave (Fig. 2-4) being most common for fiberglassing work. The plain weave, which is the most common, will usually suffice for fiberglassing repair work. There are certain jobs, though, where the smooth surface of the long shaft satin and the greater reinforcing strength in one direction of the unidirectional weave are needed.

Glass fiber cloth is available not only in various weaves but also in various weights per square yard. Weights of about 4 ounces per square yard up to about 20 ounces per square yard are useful for various types of repair work. Weights from about 6 to 10 ounces will take care of most jobs. These are dry weights of the cloth, before it has been saturated with plastic resin.

When used to make fiberglass laminates, cloth gives the most strength but the least thickness. It

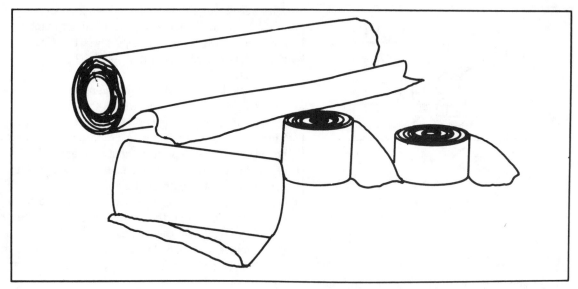

Fig. 2-1. Fiberglass cloth.

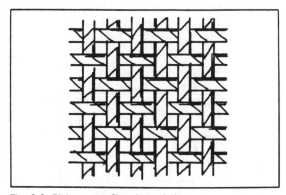

Fig. 2-2. Plain weave fiberglass cloth.

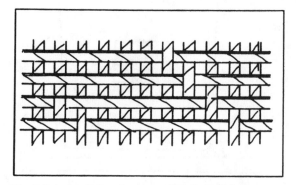

Fig. 2-3. Long shaft satin weave fiberglass cloth.

usually takes 40 to 50 layers of 12 ounces per square yard cloth to make a laminate 1 inch thick.

Cloth is available by the yard in widths from about 38 to 60 inches with selvaged edges so that it will not unravel. Cloth tape with selvaged edges is available in widths from 1 to 12 inches (Fig. 2-5).

The selvaged edges make the cloth easier to handle, so the width should be selected for the particular repair job. At the unselvaged ends and where cuts are made, the cloth tends to unravel. it must be handled carefully when dry and when plastic resin is being applied to it.

Cloth is a fairly easy reinforcing material to use on flat surfaces, but it is difficult to keep wrinkle-free in curved areas. Cuts usually have to be made to form square corners and various other

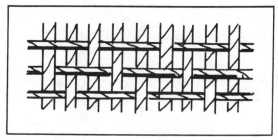

Fig. 2-4. Unidirectional weave fiberglass cloth.

9

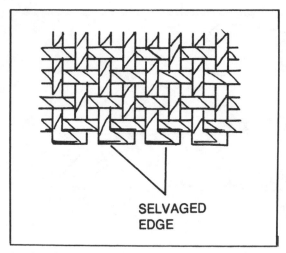

Fig. 2-5. Selvaged edge of fiberglass cloth.

configurations. Cloth gives a fairly smooth finish, especially if lighter weight cloths are used.

Cloth requires the least resin of all the reinforcing fabrics. This makes the cloth very strong, but it lacks stiffness. Plus, it does not give good waterproofness because of the limited amount of resin (Fig. 2-6). It's the resin and not the reinforcing material that makes fiberglass highly impervious to water. To get around this problem, cloth is usually combined with *mat,* which takes more resin, in laminates requiring waterproofness.

Cloth is the most expensive of the reinforcing materials on a weight basis. For many fiberglassing repair tasks, it has advantages that make it worth the additional cost.

Because untreated glass fiber cloth is also available, it's extremely important that you purchase only the kind specially treated for resin and fiberglass laminating for use in fiberglassing repair

work. Be careful about "bargains" in fiberglass cloth. It might be unchromed and thus unsuitable for fiberglassing. The chroming gives the cloth a shiny appearance; untreated fabric has a dull appearance.

Glass fiber cloth differs from most woven materials in that it has a slip weave that allows the threads to slide over each other to a certain extent. This allows the dry fabric to be shaped to compound curves in one piece without cutting it. There are definite limits to the amount of compound curving that can be accomplished.

Fiberglass cloth can be purchased by the roll or by the yard cut to the desired length from the roll. A straight cut can be made by cutting and pulling out one strand that goes across the cloth. This serves as a guideline for cutting the fabric with scissors or shears.

Glass Fiber Woven Roving

Glass fiber woven roving is shown in Fig. 2-7. Whereas cloth is made from glass fiber thread that

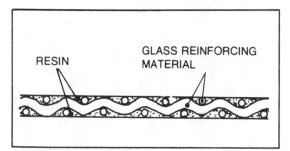

Fig. 2-6. Fiberglass cloth laminate.

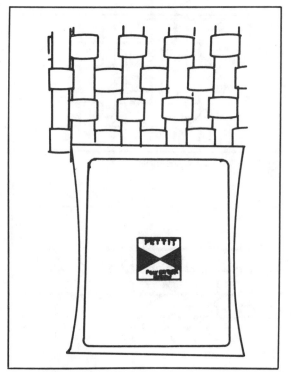

Fig. 2-7. Fiberglass woven roving.

is twisted like yarn, woven roving is made from rovings or continuous strands of glass fibers that are grouped together. Woven roving is a thick, clothlike reinforcing material.

While woven roving is available in various weights per square yard, 24 ounces per square yard (dry) is the most common. It serves for most fiberglassing repair work needing woven roving.

When used to make a fiberglass laminate, woven roving of 24 ounces to the square yard (dry) gives about an inch thickness (Fig. 2-8). In practice, woven roving is almost always alternated with layers of mat to fill in the heavy weave pattern. This combination gives good adhesion, stiffness, and tensile strength.

Woven roving is available by the yard in widths from about 38 to 60 inches. The woven roving comes with unselvaged edges. Take care when handling the woven roving, so it will not come unraveled.

Woven roving can be purchased by the roll or by the yard cut to the desired length from the roll. Care should be taken when cutting it to prevent unnecessary unraveling. Woven roving is also sometimes sold in plastic bags, usually with about a square yard of the material, but this is generally more expensive than purchasing it by the yard from a roll.

On a weight basis, woven roving is less expensive than cloth. It is more expensive than mat.

Woven roving has more strength than mat, but less than cloth. The woven roving has the advantage over cloth in that it gives a more rapid buildup of thickness. Because of the heavy basket weave, woven roving does not give as smooth a surface as cloth.

A main disadvantage of woven roving as compared to both cloth and mat is that the woven roving is more difficult to wet out and saturate with resin. Woven roving requires more resin than cloth. A woven roving laminate is generally about 45 percent glass fibers and 55 percent resin, whereas a cloth laminate is about 50 percent glass fibers and 50 percent plastic resin. Both woven roving and cloth require much less resin than mat.

Because of the relatively low resin content, woven roving, like cloth, gives a laminate that lacks waterproofness. To get around this problem, woven roving is usually combined with mat, which takes more resin, in laminates requiring waterproofness. It's the resin and not the reinforcing material that makes fiberglass highly impervious to water.

Woven roving, like cloth reinforcing material, has a slip weave. This allows the rovings to slide over each other to a certain extent. Individual strands slip over each other, so the dry fabric can be shaped to compound curves in one piece without cutting it. There are definite limits to the amount of compound curving that can be accomplished. The heavy weight and thick basket weave make woven roving much more difficult to work with than cloth.

As far as woven roving is concerned in fiberglassing repair work, there are times when the advantages outweigh the disadvantages and difficulties in working with this material. When used for repair work, the woven roving is usually sandwiched between layers of mat. This helps to prevent a resin-rich laminate, which tends to be brittle, and gives a smoother outside surface.

Glass Fiber Mat

Glass fiber mat is shown in Fig. 2-9. It is manufactured by laying down chopped strands of glass fiber in a random pattern on a flat surface. A bonding agent is used to bond the strands together into a feltlike material.

Unlike cloth and woven roving, mat is sold in weights per square foot rather than per square yard. Mat is commonly available in weights ranging from about 3/4 ounce per square foot to 3 ounces per square foot.

Mat is available in widths from about 36 to 60

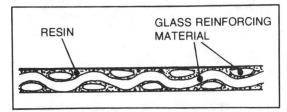

Fig. 2-8. Fiberglass woven roving laminate.

Fig. 2-9. Fiberglass mat.

inches. It can be purchased by the roll or by the yard cut from the roll. Mat is also sometimes sold in plastic bags, usually with about a square yard of the material. This can be more expensive than purchasing it by the yard from a roll.

Mat is the least expensive of the reinforcing materials on a weight basis. This advantage is partially offset by the fact that it requires more resin than cloth or woven roving. Like cloth, woven roving, and other glass fiber reinforcing materials, mat must be treated, such as by chroming, before it is suitable for use with plastic resins.

Mat has a lower glass per resin ratio, usually from 25 to 35 percent, than cloth or woven roving laminates. Because of the high proportion of resin in the mat laminate, it yields a weaker laminate than either cloth or woven roving. Mat has the advantage of forming the stiffest laminate. Also, because of its nondirectional random pattern, it has the best inner bonding strength when used in a laminate (Fig. 2-10).

Because of its high resin content, a mat laminate has the best waterproofness. It forms a much better barrier against water than either cloth or

woven roving. Mat is thus frequently used as a gasket when joining fiberglass laminates. The mat is first saturated with catalyzed resin and placed between the two moldings before they are joined together.

Mat is also useful for filling in the coarse wave of woven roving. This reduces the amount of resin without reinforcing glass fibers and increases the strength of the resulting laminate (Fig. 2-11).

For many fiberglassing repair jobs, mat is the easiest material to use. It is easy to wet out or saturate with resin. While dry mat is fairly stiff and will not shape to compound curves, when wetted out or saturated with plastic resin, the binder holding the mat together dissolves. This allows the mat to be shaped easily to compound curves and other configurations.

Care must be taken when applying wet resin to mat to keep it from bunching up and forming lumps. This is seldom much of a problem after the correct techniques for working with this material are learned.

It might appear that the short strand of chopped fibers, usually from about ¾ inch to 1½ inches in length, are responsible for making mat laminates weaker than cloth and woven roving laminates that have long continuous strands of glass fibers. Tests indicate that this has only a slight effect on the strength of the laminate. The mat laminate is weaker mainly because of the lower glass-to-resin ratio.

In many fiberglassing repair jobs, either mat or cloth can be used to achieve about the same results. In this case I recommend that mat be used,

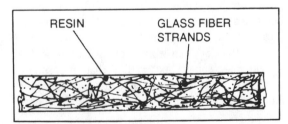

RESIN   GLASS FIBER
               STRANDS

Fig. 2-10. Fiberglass mat laminate.

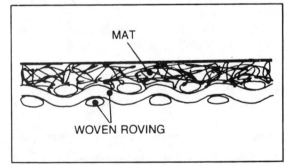

MAT

WOVEN ROVING

Fig. 2-11. Using mat to fill coarse weave pattern of woven roving.

as it is much easier to wet out or saturate with resin and, unlike when using cloth, air bubbles are seldom a problem. This point will become clear when the practice projects detailed in later chapters for learning fiberglassing techniques are carried out.

In a laminate, it takes about 20 layers of 1½-ounce-per-square-foot (dry weight) mat to build up a 1-inch thickness. Because of the higher resin content and resulting lower strength and modulus of elasticity, mat laminates must be thicker than those of cloth or woven roving to have equivalent properties.

Surfacing Mat

A special purpose mat that is formed from individual glass fibers rather than whole strands is also available. This mat has a thin body and gives a very smooth surface. It is useful for surfacing and overlay work. For example, when regular mat is used to build up a laminate, a layer of *surfacing mat* can be added to give a smooth surface. It is also used to make lampshades and other products requiring a thin smooth material. The glass fibers can be seen through the resin, giving an interesting effect.

Surfacing mat is available in thicknesses from about .01 to .03 inch in widths from about 38 to 60 inches. It is sold by the roll, which usually has about 200 yards, or by the linear yard cut from a roll. Surfacing mat is also sold in plastic packages, which usually contain about a square yard of the material. These are usually more expensive than purchasing the material by the yard from a roll.

Surfacing mat can be useful for some types of fiberglassing repair work. It also has many hobby and craft uses.

Chopped Glass Fiber Strands

Chopped strands of glass fiber rovings are another important glass fiber reinforcement for fiberglassing repair work. They are available in lengths from about ¼ to 2 inches. The glass fiber must be chrome-treated before it is suitable for fiberglassing work. Chopped strands are usually sold in packages by weight.

Chopped strands are added to resin to give structural strength. Chopped strands produce a weaker laminate than glass fiber cloth, woven roving, or mat, however. The chopped strands should be thought of as a supplement to, rather than a replacement for, these other reinforcing materials.

The chopped strands can also be combined with resin and a *thixotropic* (thickening) powder to make an excellent putty that can be sanded, ground, and drilled after it has hardened. When only a limited amount of chopped strands are needed, they can easily be pulled from regular mat.

When using chopped strands, keep the ratio of chopped strands high in relation to the amount of resin used. This will result in the greatest structural strength and will help keep the resulting material from being overly brittle because it is rich in resin.

Combination Woven Roving and Chopped Strands or Mat

Woven roving and chopped strands or mat combined into a single material are also available. This combination material is a time-saver for fast buildup of large fiberglass moldings. For most fiberglassing repair work, this reinforcing material is not needed. The main advantage is that a layer of woven roving and chopped strands or mat is laid up in one operation, which cuts the labor cost in manufacturing operations. The main disadvantages are that the material is difficult to handle and wet out or saturate with resin. It's usually much easier to first lay up a layer of one of these reinforcing materials, followed by a layer of the other material, than to use the combination material, although it will probably take longer.

Various weights are available, with about 24 ounces per square yard being typical. It is available in widths from about 38 to 60 inches. Like cloth, woven roving, and mat, it's sold by the roll or by the linear yard from a roll.

Milled Fibers

Milled fibers, sometimes called ground glass fibers, are made from glass strands. These are hammer-milled into pieces much shorter than chopped strands.

Milled fibers are usually sold in packages by weight. Some large suppliers sell them by the pound in bulk.

Milled fibers can be added to polyester or epoxy resins to form a jellylike paste that makes an excellent filler material. The consistency of the compound can be controlled by the amount of milled fibers added to the resin.

A lesser amount of milled fibers can be added to resin as a thickening material to keep the resin from running down or sagging when doing fiberglassing lay-up work on vertical or overhead surfaces. There are other thixotropic or thickening materials, described later in this chapter, that generally give better results than milled fibers.

Milled fibers mixed with resin usually do not result in as much structural strength as when chopped strands are similarly used. Thus, the milled fibers mixed with resin should be used only for filling in small areas.

NONFIBERGLASS REINFORCING MATERIALS

Several nonfiberglass reinforcing materials, including Vectra polypropylene, Dynel Modacrylic fiber, Xynote polyester, DuPont Kevlar 49 Aramid, and carbon fiber, are now available for use with polyester and epoxy resins. These nonfiberglass reinforcing materials are generally more expensive (some of them much more expensive) than those of glass fibers. For some types of fiberglassing repair work, though, they have advantages that make them worth the additional cost.

Laminates made up of these nonfiberglass reinforcing materials and plastic resin are still referred to as fiberglass even though there are no glass fibers in the material. This broad use of the term will be used in this book.

Vectra Polypropylene

Vectra polypropylene reinforcing material is made from an extremely lightweight textile fiber. Standard 4.3 ounce-per-square-yard polypropylene cloth has about the same bulk and absorption as 10-ounce fiberglass cloth. Polypropylene has a higher strength-to-weight ratio than fiberglass reinforcing material. It gives a useful combination of lightweight and high tensile strength. Other advantages include greater abrasion resistance, greater elasticity, and better bonding adhesion to wood than fiberglass reinforcements. Unlike fiberglass reinforcing materials, polypropylene does not cause skin irritation during handling and sanding and is *nonallergenic*. This is an extremely important advantage for those who are allergic to glass fiber reinforcing materials.

Disadvantages of polypropylene include higher cost (approximately a dollar a yard more than 10-ounce-per-square-yard fiberglass cloth, which has about the same bulk and absorption of resin as 4.3-ounce-per-square-yard polypropylene), a tendency to hold its creases, and a tendency to float in wet resin, making it somewhat difficult to lay down without the formation of air bubbles. Methods for minimizing the problems with creases and laying down the fabric are detailed in later chapters. The high abrasion resistance of moldings laid up with polypropylene make it resistive to sanding (the polypropylene laminate tends to sand the sandpaper).

Polypropylene is easy to work with in many ways. The greater elasticity allows easy application around compound curves. The material wets out rapidly and easily.

The greater elasticity of polypropylene makes it better suited than fiberglass reinforcement for surfaces that work. To avoid cracking of the resin, special flexible polyester or epoxy resin (readily available) is recommended.

Polypropylene cloth reinforcing material is available by the roll or by the linear yard from rolls in widths from 38 to 60 inches. Fabric count is 21 × 21 in a spun plied yarn construction. Large openings in weave allow for penetration of resin during lay-up. The cloth has selvaged edges.

Polypropylene is also available in tape form in widths from 2 to 12 inches. "Tape" is narrow strips of material similar to the cloth fabric; there is no adhesive backing. The edges of the polypropylene tape presently available are not selvaged. They are machine cut. Care must be taken in handling them so that they do not become unraveled. The tape is useful in repair work for reinforcing seams and other high-stress areas.

Dynel Acrylic

Dynel is an acrylic reinforcing fabric. The acrylic yarn is made in Japan, and the finished yarn is woven into fabric in the United States. At the time of this writing, it is available only in 4-ounce-per-square-yard dry weight in a 63-inch width. It is sold by the roll or by the linear yard cut from rolls. It is presently priced slightly higher than the polypropylene.

Dynel acrylic fabric weighs only about half as much as fiberglass cloth of the same thickness. The main advantages of the acrylic, as compared to fiberglass, are greater abrasion resistance, higher tensile strength, and the fact that it does not cause skin irritation when handled or sanded. The acrylic fabric is fairly easy to work with; it is easier to stretch around sharp corners and curves than either fiberglass or polypropylene reinforcing fabric. It provides a slick finish when sanded, making it ideal for use as an overlay. A canvaslike finish is possible by using a minimum of resin. In either case, the fabric is easy to wet out. The main use of Dynel acrylic fabric is as an overlay for a fiberglass laminate to provide greater abrasion resistance. The acrylic fabric provides much better adhesion to wood than fiberglass, although not quite as good as the polypropylene fabric described earlier. Both polyester and epoxy resins can be used.

Xynole Polyester

Xynole polyester is an ultrafast wet-out reinforcing and overlay fabric. It is generally considered to be the easiest reinforcing material to apply. It conforms easily to compound curves and wets out easily and rapidly without leaving air bubbles.

Other advantages of Xynole polyester, as compared to fiberglass, are that Xynole polyester does not cause skin irritation from handling or sanding and has better laminating adhesion with either polyester or epoxy resin. Also Xynole polyester has greater toughness and abrasion resistance and better weight-to-strength ratio.

The main disadvantage of Xynole polyester is that it lacks stiffness. For this reason, it is usually not used alone for laminating in areas where high stresses will exist. It makes an ideal overlay for

fiberglass reinforced laminate and as a waterproofing coating for plywood and metal.

Xynole polyester is available in a 4.2-ounce-per-square-yard dry weight cloth, which has about the same bulk and absorption of resin as 10-ounce-per-square-yard dry weight fiberglass reinforcing cloth. It is available by the roll or by the linear yard cut from rolls in widths from 38 to 60 inches. Xynole polyester is presently priced only slightly higher than 10-ounce-per-square-yard fiberglass cloth. It is also available in tape form in widths from 1½ to 12 inches.

DuPont Kevlar Reinforcing Material

Kevlar is a synthetic, long-chain, polymer fiber developed by DuPont in 1972. It is now available in a 5-ounce-per-square-yard reinforcing fabric in 38- and 50-inch widths. This material can be used with polyester or epoxy resin. Laminates from this material have greater tensile strength than when fiberglass reinforcement is used, but they are not as strong in compression or bending. An important advantage is superior impact strength. A big disadvantage of this material for fiberglassing repair work is that it is much more expensive than fiberglass reinforcing material.

Carbon Fiber Reinforcing Material

This is another reinforcing material that is used with epoxy resin for laminating purposes. It is expensive. Carbon fiber reinforcing fiber is used for some custom boat construction where high strength with minimum weight is desired.

SOME REINFORCING MATERIALS FOR REPAIR WORK

For most of the fiberglassing repair work detailed in this book, the less expensive glass fiber reinforcing materials can be used. In the few cases where the more expensive nonfiberglass reinforcing materials are called for, the least expensive material that will give satisfactory results is recommended.

15

RESINS

While there are many types of resins that can be combined with reinforcing materials to form reinforced plastic (commonly called fiberglass), the concern here will be with only the two main ones: *polyester* and *epoxy*. All of the fiberglassing repair work detailed in this book can be accomplished with these resins. Other types such as silicone, phenolic, melamine, and thermoplastic resins, while useful in some manufacturing processes, will not be required.

Polyester Resins

Polyester resin is a thermosetting plastic. *Thermosetting* means that it is set or cured by heat, which can either be applied chemically from inside the resin (*exothermic heat*) or outside, or that a combination of the two can be used (Fig. 2-12). An *accelerator* (a highly active oxidizing material) and a *catalyst* are added to the liquid polyester resin to start the chemical reaction that causes the internal heat. The heat is the setting agent. When polyester resin is in liquid form, the molecules lay side-by-side in no set pattern. The addition of heat causes the molecules to link together in chains to form a solid plastic—a hard mass that cannot be softened by application of heat. The process of changing from a liquid to a solid is called *polymerization*.

Most of the polyester resins that are manufactured for room temperature cures, the ones that will be used for most fiberglassing repair work, already have the necessary accelerator added to the liquid resin. In most cases the accelerator is cobalt naphthenate. Only a catalyst, usually methyl-ethyl-ketone (MEK) peroxide, needs be added to the polyester resin to start the cure at room temperature.

Polyester resin will harden at room temperature in time even without the addition of the catalyst. With the catalyst, however, the resin hardens quickly, usually in about five minutes to an hour, depending on how much catalyst is added, room temperature, humidity, and other factors. The thickness of the layer of resin applied also affects the rate of cure. A thick layer keeps the heat inside better, causing it to cure faster than a thin layer.

The fact that polyester resin will harden in time at room temperature means that liquid polyester resin, even when in sealed containers, has a limited shelf life. When purchasing resin, always make sure that it is fresh and has not started to cure in the container.

Fig. 2-12. Polyester resin patching kit (courtesy Pettit Paint Company, Inc.).

16

Laminating and Finishing Polyester Resins

Many types of polyester resin have been developed for specific purposes. The two basic types most commonly used in fiberglassing repair work are: *laminating* or *lay-up* and *finishing* or *surfacing*.

The laminating or lay-up resin is *air-inhibited,* which means that in the presence of air it will not fully cure. The surface will remain tacky. This condition is desirable when additional layers of fiberglass are to be added to the laminate, as there is no waxy surface to prevent laminating additional resin layers together properly.

Finishing or surfacing resin is *nonair-inhibited,* which means that it will fully cure in the presence of air. This is desirable for final application to a laminate when a complete cure is wanted. Nonair-inhibited resin has a wax or similar ingredient added. When the catalyzed resin is added to a laminate, the wax rises to the surface, sealing off the air and allowing a complete cure. The surface can then be sanded (the tacky surface that results when air-inhibited resin is used will quickly gum up sandpaper).

Laminating or lay-up polyester resin can be made to cure tack-free by adding a special wax to the liquid resin before application. Still another method of achieving a surface cure when laminating or lay-up resin is used is to seal the surface from the air. One way to do this is to cover the surface with cellophane, taping it in place around the edges or otherwise securing it so that air cannot get below.

When finishing or surface resin is used, it cures with a waxy surface. If for any reason another layer of laminate must be added to this, the wax should first be removed. This can be accomplished by sanding or using a solvent such as *acetone*.

A general-purpose resin that can be used for both laminating or lay-up and finishing or surfacing work is also available. Some fiberglass workers look down at this resin, saying that it doesn't do either job well. I have achieved satisfactory results with it for many types of fiberglassing repair work — that is, provided that I used quality brands. For large jobs, I highly recommend the use of both laminating and finishing resins rather than a general-purpose resin.

High-Viscosity Polyester Resins

Polyester resins are available with either regular or high viscosity. The high-viscosity resins contain a thixotropic or thickening agent that helps eliminate dripping and sagging on vertical and overhead surfaces. Special thixotropic powder is added to regular resin gives it a higher viscosity. Some types actually strengthen the resin. Avoid the use of clay and inert earth as a thixotropic agent as these will weaken the cured resin. Thixotropic powder suitable for both polyester and epoxy resins are sold in bags or containers by weight or in bulk by weight.

Flexible Polyester Resins

Polyester resins are available that have various degrees of flexibility when cured. Some are very rigid when cured; others are quite flexible. Most are somewhere between these extremes. Special flexible polyester resin is recommended for use with the nonfiberglass reinforcing materials that stretch or elongate more than fiberglass reinforcing materials do. As compared to more rigid resins, flexible ones result in a laminate that will crack and craze less from impact. More rigid resins should be used when greater stiffness is desired.

Fire-Retardant Polyester Resin

Cured laminates made up of regular polyester resin will support combustion about the same as plywood. Special fire-retardant polyester resin is available. It is more expensive than the regular resin, but this extra cost is warranted for use in constructions and repairs requiring fire retardancy. The fire retardancy is achieved by a coreactant compound that is combined with the liquid resin. This results in a slower curing time for the resin, but adhesion, flexibility, and impact resistance are not affected. The coreactant adds a white opacity to the resin, and the usual translucency is lost.

Polyester Resins and Temperature

Most polyester resins are formulated for use in a temperature range from about 70 to 75 degrees Fahrenheit. Satisfactory results can be achieved in a temperature range from 60 to 70 degrees Fahrenheit by adding additional catalyst, and by adding less catalyst in a temperature range from 75 to 90 degrees Fahrenheit. This will be explained in detail later. For now it's important to realize that regular polyester resin is suitable for use in a Fahrenheit temperature range of 60 to 90 degrees, with difficulties increasing near the extremes.

Special low-temperature resin has been specially formulated for use in a temperature range from about 45 to 60 degrees Fahrenheit. This allows satisfactory fiberglassing repair work to be done at these colder temperatures. This is a *preaccelerated resin* (an accelerator has already been added by the manufacturer to the liquid resin) that is catalyzed in the same manner as regular polyester resin. The shelf life of low-temperature resin is limited to three or four months, and the entire contents of the container should be; used by that time.

Gel Coat Polyester Resin

Still another type of polyester resin is gel coat. Its main use is to form the protective color gel coat surface on a fiberglass molding. It is available in clear form for adding your own pigment or with the color added. Most gel coat resin is air-inhibited. It is applied over a mold release agent in a female mold. The lay-up of resin and reinforcing materials follows. The gel coat will form the outside color surface on the finished molding. The gel coat cures because it is air-inhibited against the mold. It will also cure when sealed off from the air by wax or other covering. Gel coat resin is specially formulated to give a protective waterproof color coat.

For touching up small areas of gel coat when doing fiberglass repair work, it is usually easier to use nonair-inhibited resin so that it will cure in air. Color pigments, available in liquid, paste, and powder forms, can be used with polyester or epoxy resin. The paste form is generally easiest to mix and blend.

Special gel coat spray kits are available for gel-coat touch-up work. Kits are also available for mixing and matching colors. Methods for applying the gel coat for repair work include knife, brush, and spray.

Polyester gel coat resin is catalyzed in the same manner as regular polyester resin. Because the gel coat forms a layer without reinforcement inside, it must be a thin layer if cracking and crazing are to be prevented.

Water Clear Polyester Casting Resin

Another type of resin that has some uses in fiberglassing repair work and many craft applications is water clear polyester casting resin. This resin is well suited for encapsulating biological and geological specimens, making decorative castings, and for use on surfboards. Either opaque or transparent color pastes can be added to this resin. The resin is catalyzed in the same manner as is regular polyester resin.

Containers for Polyester Resins

Polyester resins are available in many quantities (pint, quart, gallon, and 5-gallon and 55-gallon drums) in various types of containers. Because the average life of regular polyester laminating and finishing resins, when properly stored in a cool place, is only about six months to a year, it is extremely important that you buy only fresh resin. One way to help assure getting fresh resin is to buy from a volume dealer.

Purchase only the amount of resin that you plan to use in a reasonable amount of time and store it in a cool place. This is especially important in the case of special-purpose resins, which often have an even shorter shelf life than regular polyester laminating and finishing resins.

Use quality brands. The difference between working with quality resin and a poor grade is considerable. Quality brands are generally worth the additional cost.

Catalyst

The catalyst, methyl ethyl ketone (MEK) peroxide, is sold in various types of containers. These are usually clear plastic so that you can see the amount of catalyst inside. There is a quantity scale marked on the side of the container. Often the container is constructed so that the catalyst can be released by drops. While the amount of catalyst to be added depends on the working temperature, the amount of resin to be catalyzed, and the amount of pot or working time desired (time until the resin starts to gel or set up), it is a small amount in relation to the volume of resin. Usually about ½ percent of catalyst by volume will give a working life of about 45 minutes at 70 degrees Fahrenheit. If only a small amount of resin is to be catalyzed, it's extremely important to have an accurate and fast method for measuring and adding a very small amount of catalyst to the resin.

Sometimes sufficient catalyst comes with the resin; in other cases it must be purchased separately. In either case, make certain that you have enough for the temperature conditions in which you will be working.

Styrene

Styrene, readily available from fiberglass suppliers, is a coactive thinner for polyester resins. Up to 40 percent by volume can be added without harming the stable character of the resin during curing. Thinning, however, is usually necessary only if the resin is to be applied by a spray gun. This is not recommended unless you have the proper safety and protective equipment (see Chapter 4). For most fiberglassing repair work, spraying resin is not necessary.

Special bubble inhibitor additives are available that can be used with either polyester or epoxy resin. When added to the resin in recommended amount, bubble formation is suppressed. The additive does not affect the curing of the resin.

Polyester Resin Filler, Putty, and Patching Compounds

A variety of polyester resin filler, putty, and patching compounds are now on the market (Figs. 2-13 and 2-14). Most of these compounds can be

Fig. 2-13. Polyester resin filler (courtesy Pettit Paint Company, Inc.).

used with or without additional reinforcing material. Various materials are used in their manufacture for thickening and strengthening the resin. Polyester resin putties are formulated for special purposes such as for boats or auto body repair. Those putties for boats should not have any additives that will rust or corrode. Those putties for auto body repair must bond well and be flexible, so they will not crack when hard.

A catalyst is added to these compounds to start the curing or hardening process. Often this is the same liquid catalyst that is used for regular polyester resins, but some polyester resin body filling compounds use a paste catalyst.

Fig. 2-14. Polyester repair putty (courtesy Rule Industries, Inc.).

Most of these compounds have a putty consistency that makes them easy to work with. The manufacturer's directions for adding catalyst and mixing should be followed carefully. Most can be applied with a putty knife, and they will stay in place until they have cured. These generally have some shrinkage, so this must be taken into account in their application.

Acetone

Acetone can be used for cleaning uncured polyester resin from brushes and tools. This must be done before the resin has set up or hardened. Do not, however, use acetone as a thinning agent for polyester resin. It will evaporate rather than participate in the cure, causing shrinkage. Even though I have seen acetone widely used for hand cleaning in manufacturing plants, this is definitely not recommended from a health and safety standpoint (see Chapter 4).

Epoxy Resins

The second main type of plastic resin used in fiberglassing repair work is epoxy. Like polyester resin, it is a thermosetting plastic (Fig. 2-15).

Fig. 2-15. Epoxy resin patching kit (courtesy Pettit Paint Company, Inc.).

The viscosity of epoxy resins, which are produced by the reaction of epichlorohydrin and bisphenal-A, varies from a thin liquid to a thick paste. These have varying molecular weights.

Unlike polyester resins, epoxy resins use a curing agent or hardener, rather than a catalyst, that actually enters into the reaction. The volume of cured epoxy resin will be approximately equal to the combined volume of the epoxy resin and the curing agent or hardener that was added to it.

As compared to polyester resin, epoxy resin has many advantages. It has greater strength and adhesion. Unlike polyester resins, epoxy resins will give strong adhesion to hardwood, metals, and glass. Epoxy resin also has lower shrinkage on curing, an important advantage in some types of fiberglassing repair work. Cured epoxy resin has even better chemical resistance than polyester resin, which has good chemical resistance.

In comparison to polyester resin, epoxy resin has many disadvantages. Epoxy is more expensive, costing more than twice as much as polyester resin. For this reason, epoxy resin is used in fiberglassing repair work only when its special properties are required for a particular job or part of a job. For example, epoxy resin can be used for bonding the first layer of a fiberglass laminate to wood or metal. Once this has cured, additional layers can be added to the laminate with less expensive polyester resin. Epoxy and polyester resins are usually compatible if one is allowed to cure before the application of the other.

Another disadvantage of epoxy resin is that it is more difficult to use than polyester resin. Epoxy requires more time and higher temperature for curing. The formulation of modern epoxy resins has been improved to the point where they handle in a manner similar to polyester resin, except for the higher required temperature and longer required curing time. Heat lamps can be used to accelerate the curing of epoxy resins.

Perhaps the biggest disadvantage of epoxy, at least from the point of view of the person using it, is the potential health hazard. While the cured epoxy resin is innocuous physiologically, this is certainly not true when the epoxy and curing agent are in liquid form. Contact with the skin can cause severe dermatitis and other problems. Exposure to the vapors can cause skin irritations, sensitization, and other problems.

When using epoxy resins, it is important that strict health and safety precautions be observed (see Chapter 4). Proper storage of epoxy resins is also important, as there is the potential danger of spontaneous combustion.

Epoxy Resins and Temperature

Epoxy resin comes in two parts, each in a separate container: the basic resin and the curing agent or hardener (Fig. 2-16). The ratio of curing agent or hardener to resin varies depending on the brand and type used. One popular type, for example, uses 4 parts by volume of resin to 1 part by volume of curing agent; another type uses 6 parts by volume of resin to 1 part by volume of hardener; and still another uses a 50-50 mix. The exact amount depends on the desired pot life and, to a certain extent, on the working temperature. The

Fig. 2-16. Epoxy resin (courtesy Rule Industries, Inc.).

best working temperature for epoxy resin is about 80 to 85 degrees Fahrenheit, which is higher than the typical 70 to 75 degrees Fahrenheit for polyester resin. While epoxy resin has a pot life of from 10 minutes to 2 hours, depending on the amount of curing agent or hardener added and other factors, it takes much longer for the resin to cure when applied to a surface. The atmosphere cools the surface, reducing the heat that is generated in the mixture. The amount of hardener that results in a pot life of about 30 minutes gives a curing time of about 20 hours at 70 degrees Fahrenheit when applied to a surface, by means of a heat lamp, can reduce the curing time of the same mixture to 2 to 3 hours.

Curing Agent and Thinner for Epoxy Resin

A special flexible curing agent is available for use with epoxy resin. This will result in a cured resin that has greater flexibility than when a regular curing agent is used.

Special epoxy thinner is available that reduces viscosity without decreasing strength or tenacious adhesion. Up to 40 percent by volume may be added without affecting the amount of curing agent to be used or the curing time. A special epoxy solvent is available for cleaning uncured epoxy resin from brushes and tools.

The same color pigments and thixotropic thickening powders described earlier for polyester resins can also be used with epoxy resins. Epoxy resin can be used with any of the fiberglass and nonfiberglass reinforcing materials described earlier in this chapter.

Epoxy Resin Filler, Putty, and Patching Compounds

Many epoxy resin filler, putty, and patching compounds are now on the market (Fig. 2-17). Most of these can be used with or without additional reinforcing materials. Epoxy resin putties are formulated for many uses. There is even a type that cures under water. These putties generally come with the necessary curing agent or hardener. Many types cure rapidly, with a working time of about 10 minutes to an hour, depending on the

Fig. 2-17. Wet or dry cure epoxy repair putty (courtesy Rule Industries, Inc.).

particular brand, ratio of curing agent or hardener and resin putty compound, and other factors. Because some brands contain metal particles, it's important that these be of a noncorroding metal alloy if they are to be used for marine applications or other jobs where rust and corrosion will be a problem.

When selecting epoxy resins and compounds, remember that there are substantial differences in the qualities of various brands. The difference between working with a quality epoxy product and one of poor quality can be considerable. Quality brands are worth the additional cost. Once you find a product that works well for your particular needs, it's best to stick with it.

Unlike polyester resin, epoxy resins can be formulated for use without reinforcing materials as glues, sealers, and paints. Epoxy is well known for its outstanding adhesion and bonding properties (Fig. 2-18).

CORE AND FORMING MATERIALS

There are many core and forming materials that are useful in fiberglassing repair work. These become part of the structure of the finished fiberglass repair or construction. Molds usually do not end up as part of the finished product. The core and forming materials might or might not add struc-

tural strength to the fiberglass laminate in themselves. Sometimes they serve merely as nonstructural cores for shaping fiberglass stiffeners and other reinforcing members.

Phenolic Microballoons

Phenolic microballoons are microscopic hollow balloons that are combined with either polyester or epoxy resin to form a lightweight material that can be troweled in place. The cells of the phenolic microballoons are filled with nitrogen. The material has a density of about 10 pounds per cubic foot. Even when mixed with resin, the resulting core material will float.

Phenolic microballoons have a flourlike consistency and are red in color. They are sold in packages by weight or by the bulk by weight. They are fairly easy to use and form an ideal core material when weight is to be held to a minimum. One disadvantage is the fairly high cost of phenolic microballoons, presently over $5 a pound.

Fig. 2-18. Epoxy glue (courtesy Pettit Paint Company, Inc.).

22

Honeycomb Cell Paper

A special honeycomb cell paper is available for fiberglass sandwich construction. It is used to form a honeycomb structure that is light in weight between two layers of fiberglass. The honeycomb cell paper contours to compound curves. It can be useful for some types of fiberglassing repair work.

Honeycomb cores can be formed from aluminum, fiberglass, and other materials. These cores are difficult to fabricate and bond to fiberglass laminates. While the honeycomb cores find some use in manufactured products, they are rarely needed for typical fiberglassing repair work.

Foamed Plastic

Polystyrene, polyurethane, and polyvinyl chloride (PVC) in foamed plastic form are frequently used as core and forming materials in fiberglassing repair work. These foams are available in various densities, with those from 6 to 8 pounds per cubic foot used most for fiberglassing repair work. Polystyrene and polyurethane foamed plastics are also used in fiberglass construction of boats, docks, rafts, and other related things as a buoyancy material. The foams used for this purpose usually have about 2 pounds per cubic foot density.

Polystyrene foamed plastic is available in precured blocks and sheets in various densities. It is usually less expensive than polyurethane or polyvinyl chloride foamed plastic, but it is also less resistant to water, decay, and damage from impact. Also, polystyrene is attacked by polyester resins, so it cannot be used as a core or forming material when polyester resins are used. It will work with the more expensive epoxy resins.

Polyurethane foamed plastic is available in precured blocks and sheets in various densities and in foam-in-place liquids, the latter formed by mixing polyol and toluene diisocyanate (TDI) together. The reaction causes the components to foam and cure into polyurethane foamed plastic. Foam-in-place and spray-in-place kits. The cure of this foam is sensitive to the ratio of the components, temperature variations, and other factors. Follow the manufacturer's directions carefully when using foam-in-place polyurethane.

Polyurethane foamed plastic is generally more resistant to water, decay, and damage from impact than is polystyrene foamed plastic. Unlike polystyrene foamed plastic, polyurethane can be used with polyester resin. Neither of these plastic materials, once cured to rigid form, can be bent to conform to compound surface curvature. They can be shaped by sawing, filing, and rasping. Foam-in-place polyurethane can be used to fill compartments and cavities of most any shapes and sizes.

Polyurethane foamed plastic is generally more expensive than polystyrene foamed plastic. The polyurethane foamed plastic has advantages that frequently make it worth the additional cost for fiberglassing repair work.

Polyvinyl chloride (PVC) foamed plastic is a closed cell material that is frequently used as a core in fiberglass sandwich construction, especially for boat hulls, decks, and superstructures (Fig. 2-19). It is available in a thermoplastic form. By application of heat, it can be formed into compound shapes. It can be bent in simple curves at room temperature without breaking down cell structures.

Polyvinyl chloride (PVC) foamed plastic is more expensive than polystyrene and polyurethane foamed plastics, but it offers many advantages that are important for certain applications. Because of its closed cell structure, it will not absorb water. It has better chemical resistance than either polystyrene or polyurethane and is nonaging. Polyvinyl chloride foamed plastic will not become brittle, crumble, or deteriorate — common problems with polystyrene and, to a lesser degree, polyurethane

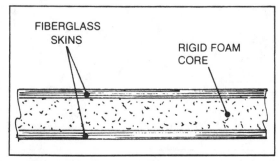

Fig. 2-19. Fiberglass sandwich construction with rigid foam core.

foamed plastics. Polyvinyl chloride foamed plastic will accept either polyester or epoxy resin. It is available in ¼, ⅜, ½, and ⅝-inch thick sheets 36 inches by 72 inches in size.

A popular use of polyvinyl chloride foamed plastic is as a planking material shaped over a male plug (Fig. 2-20). A fiberglass skin is then laminated to the foamed plastic. The male plug is then removed, and the other side of the foamed plastic is sheathed with a fiberglass laminate. The result is a fiberglass laminate with a polyvinyl chloride foamed plastic core in sandwich construction. This

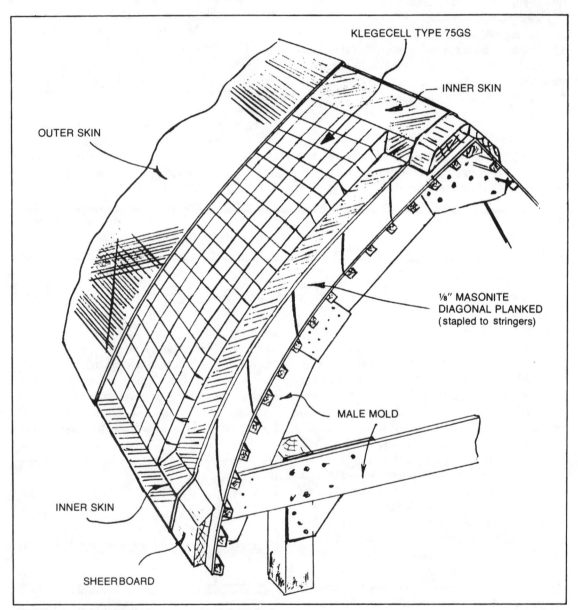

Fig. 2-20. Use of Klege-Cell polyvinyl chloride rigid foam for one-off boat hull construction (courtesy American Klegecell Corporation).

results in a stronger laminate than if the two fiberglass skins are laminated together into a single unit without the core material.

This same principle can be used for making many fiberglassing repairs. It is also a popular method for one-off boat construction.

Wood

Wood has frequently been used over the years as a core and stiffening material for fiberglass construction. The advantages and disadvantages of using wood for this purpose are frequently debated by fiberglassing experts. The fact that wood and fiberglass have different physical and chemical properties creates many problems when using composite wood and fiberglass construction.

Thin layers of end-grain balsa are sometimes used as a core material for fiberglass sandwich construction, especially for boats. The end-grain structure is used so that the resin can better penetrate the balsa. End-grain balsa is available for fiberglassing construction in the form of thin end-grain blocks held together in sheet form with a gauzelike backing (Fig. 2-21). This allows the balsa to be contoured to complex shapes inside a mold or over a form. Balsa is sometimes used as a core material for fiberglassing repair work.

Plywood is frequently used as a structural core material or structural reinforcement for fiberglass, especially for boat decks and superstructures (Fig. 2-22). The plywood is sometimes bonded to the fiberglass on one side only. In other cases, both sides of the plywood are sheathed with fiberglass laminates in a sandwich construction form (Fig. 2-23). Unlike foamed plastic cores, the plywood

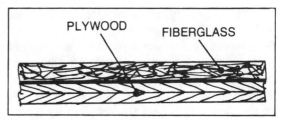

Fig. 2-22. Plywood structural reinforcement for fiberglass molding.

itself adds considerably to the strength and stiffness of the finished laminate.

Plywood can be used similarly for some types of fiberglassing repair work. Only exterior grades of plywood should be used. A common difficulty is in bonding the first layer of a fiberglass laminate to the plywood. Because polyester resin does not work well for this epoxy resin is often used for bonding the first layer of the fiberglass laminate to the plywood. After the epoxy has cured, polyester resin can be used for the remainder of the lay-up. The epoxy resin also provides greater protection from water and moisture getting through to the plywood. If the plywood does become saturated with fresh water, dry rot can become a problem.

Various softwood cores are also used in fiberglassing construction and repair work. Although epoxy resin provides greater bonding strength to the wood, the less expensive polyester resin will often suffice. In most cases, fairly substantial fiberglass sheathing laminates are required to fully protect the wood from moisture (and even then some will probably get through eventually) and keep wood swelling and cracking from damaging the fiberglass laminate surrounding the wood.

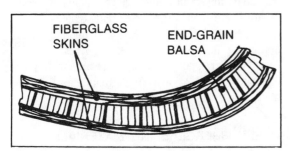

Fig. 2-21. End-grain balsa used as core material for fiberglass sandwich construction.

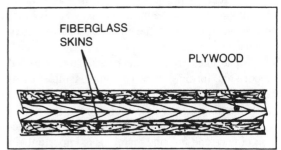

Fig. 2-23. Fiberglass sandwich construction with plywood core.

The possibility exists for treating the wood with a wood preservative to help prevent dry rot from forming in the wood sheathed in fiberglass. Most wood preservatives make it difficult or impossible to get a good fiberglass bond to the wood. They are generally not used when the wood is to be sheathed in fiberglass. Opinions do vary on this point, however, so you might want to do some experimenting to see if you can get an adequate bond for your purposes on wood that has been treated with a preservative.

Various hardwood cores are also used in fiberglassing construction and repair work. It is generally more difficult to bond fiberglass laminates to hardwood than to softwood with polyester resin. The more expensive epoxy resin will usually give a good bond. When hardwood is sheathed with fiberglass as a core or structural beam or other member, thick laminates should be used.

Wood is also often used to strengthen, reinforce, and stiffen fiberglass by means of bulkheads and stringers in fiberglass boat hulls. Wood can be used similarly for fiberglassing repair jobs.

When selecting wood for fiberglassing work, make certain that it is dry. Techniques for using wood in fiberglassing repair work are covered in later chapters.

Other Possible Core and Forming Materials

The core and forming materials described earlier are the most commonly used ones, but there are many other possibilities including metal, ferrocement, and plastics that are compatible with the resin being used. Polyester resin usually does not give adequate bonding strength with these materials; in most cases the more expensive epoxy materials; in most cases the more expensive epoxy resin will be required.

Sometimes even cardboard and paper are used as core and forming materials. For example, cardboard mailing tubes can be used as a core and form for laminating a fiberglass tube around it.

For fiberglassing repair work, careful selection of core and forming materials is important. In most cases, plastic core and forming materials are the most compatible with fiberglassing construc-

tion. If metal is used, it might bleed rust through the fiberglass laminate. Wood might rot and reduce the strength of the construction. Bonding strength must also be considered. In sandwich construction, for example, the shear strength of the laminate will be greatly reduced if the core material does not bond adequately to the fiberglass skins.

MOLD RELEASE AGENT

A special mold release agent is available. A thin film of mold release is applied to the mold before application of gel coat and lay-up of the fiberglass laminate, so the cured laminate can be removed from the mold without sticking. Mold release agent is also useful for many fiberglassing repair jobs. For example, it can be applied to a temporary backing, which can be wood, cardboard, metal, or plastic, that is used behind a hole to be filled in on a damaged fiberglass laminate. This will allow easy removal of the backing piece after the fiberglass laminate has cured.

FLEXIBLE MOLD COMPOUND

Flexible mold compound is another product that can be useful for some types of fiberglassing repair work and for craft projects. One type is used by applying mold release agent or a 50 percent solution of Vaseline petroleum jelly and kerosene to the object that a mold is to be made from. Flexible mold compound is then brushed or poured on in successive layers until it is about ¼ inch thick. The compound cures in about 12 hours. The mold has a hard rubbery texture.

The flexible mold compound is useful for making molds for small items. Larger molds are commonly made from other materials, such as fiberglass or wood. The materials and techniques for doing this are detailed in later chapters.

Two-part epoxy and two-part brushable polyurethane (Fig. 2-24) are popular paints for finishing fiberglass.

OTHER FIBERGLASSING MATERIALS

The materials described earlier are the ones most commonly used in fiberglassing repair work.

Fig. 2-24. Two-part brushable polyurethane finish comes with application guide (courtesy Pettit Paint Company, Inc.).

For some types of repair work, though, additional materials are required, such as sealing compounds and bolts, screws, rivets, and other mechanical fasteners. These materials will be described in later chapters along with the techniques for using them.

USE OF MATERIALS

Because materials account for a large part of the cost of fiberglassing repair work when you provide your own labor, careful shopping is important. Materials should be purchased and used in such a way that waste is kept to a minimum. Materials that end up as scraps that cannot be used should be reduced to the smallest amount that is practical. Buy in sizes and quantities that will result in a minimum of waste. While it is almost inevitable that some materials will be ruined or wasted, such as resin that accidentally is allowed to harden in the container or is spilled, strive to keep this to a minimum.

Liquids present special problems. Keep lids tight. Store liquids properly to prevent hardening in containers. Always store polyester resin in cool locations. Heat acts as a catalyst, causing the resin to start hardening. Do not store polyester resin in clear glass or plastic containers for long periods of time as ultraviolet rays act as a mild catalyst.

QUALITY OF MATERIALS

The best economy is to purchase quality reinforcing materials, resins, and other materials needed for fiberglassing repairs. Be wary of bargain materials. I've tried some of these, usually with less than satisfactory results.

All reinforcing materials should be clean when you purchase them. Avoid carelessly stored and handled materials. Make certain that resins have been properly stored and are fresh—most have shelf lives of a few months to about a year. Stick to name brands sold by reputable manufacturers.

WHERE TO PURCHASE MATERIALS

It will be most convenient, though not necessarily least expensive, to purchase fiberglassing materials in the area where you live. Provided, that is, if the stores and supply centers have what you need.

Fiberglassing materials are often sold by paint stores, auto stores, recreational vehicle supply centers, hardware outlets, large discount stores, building suppliers, and marine and boating stores. The selection will vary widely. Some areas have stores that specialize in fiberglassing materials and supplies. The one in my area is called TAP Plastics, Inc. They have a large selection of quality fiberglassing materials available at reasonable prices. I've used their products for a variety of fiberglassing repair work with excellent results. They have stores in many California cities and also in Portland, Oregon, and Salt Lake City, Utah. For information, write to TAP Plastics, Inc., 3011 Alvarado Street, San Leandro, CA 94577. Other fiberglassing material supply stores in various parts of the United States will give expert advice about fiberglassing and using their products.

I've found that some large paint store chains

offer fiberglassing materials at discount prices. I've tried some of these materials. In some cases they were of passable quality; in other cases they were difficult to work with or gave unsatisfactory results. For those who are just starting out in fiberglassing work, I don't recommend taking a chance with these low-price products.

Auto chain and discount stores frequently offer fiberglassing materials intended primarily for auto body repair and customizing. These are usually sold at reasonable prices, and I've had good results with many of these products when used for their intended purposes.

Marine stores frequently stock quality brands of fiberglassing materials. I've found their prices to be extremely high, with few exceptions. Sometimes the same brands and materials can be purchased in the same area at nonmarine stores at much lower prices. At least I have found this to be the case in many areas where I've lived.

In my opinion, the best mail-order source for fiberglassing materials, with one of the largest inventories and selections available anywhere, is Defender Industries, Inc., 255 Main St., New Rochelle, NY 10801. They stock most of the materials described in this chapter, including many of the hard to find nonfiberglass reinforcing materials. Their materials are high-quality name brands that are sold at discount prices. I've found that even with the cost of shipping figured in, the cost is generally much less than if I purchase the materials locally, assuming that I can even find what I need. This often offsets the inconvenience of having to order the materials and wait for delivery, although I've found that materials shipped by United Parcel Service arrive very quickly.

Defender Industries, Inc., has a 168-page catalog that sells for $1. In addition to the fiberglassing materials, they offer a complete line of marine products. There are also many other mail-order sources for fiberglassing supplies (see Appendix).

Another possible source of fiberglassing materials is through boatbuilding clubs or groups, which now exist in many parts of the United States. Members can often purchase materials at discount prices.

There is a limit to the amount of trouble to which you will want to go for saving money on small quantities of materials. The convenience of purchasing them locally offsets the typically higher prices you will have to pay. For large quantities of materials, it will probably be worth the trouble to shop for the lowest prices.

When I find a brand of a certain material that works well, I try to stick with it. I find, for example, that each brand of polyester resin intended for the same purpose, such as laminating or lay-up, has slightly different handling characteristics. By staying with one brand, I can better learn how to work with it and predict the results. It will take some experimenting to find brands that seem to work best for you.

Chapter 3

Tools, Equipment, Supplies, and Work Areas

The tools, equipment, supplies, and work areas required for fiberglassing repair are generally less expensive than those needed for other types of repair work. Additional tools and better work areas will enable you to do the work faster with less effort on your part. Most people start with the essential tools, equipment, and supplies, and then gradually add to them as needs develop. At first the work can be done in a garage, yard, or other available space, but later an improved work area might be needed.

The concern here will be with tools, equipment, supplies, and work areas for accomplishing fiberglassing repair work and for the protection, health, and safety of those doing the work. The tools required for specific fiberglassing repair jobs vary greatly, and not all of them are required for an individual job. Body and fender tools, for example, are almost essential for making some types of fiberglassing repairs on metal auto bodies. They will probably not be needed for, say, repairing damage to a fiberglass shower stall. Thus, selection of tools, equipment, and supplies will depend not only on how much you can afford to spend, but also on the particular type of fiberglassing repair work you intend to do.

TOOLS, EQUIPMENT, AND SUPPLIES

For our purposes here, tools, equipment, and supplies (other than materials covered in Chapter 2) are grouped together.

Mixing Cans, Cups, Buckets, and Tubs

Mixing containers are used for mixing resins with catalysts or curing agents, for holding solvents for cleaning brushes and tools, and so on. Clean empty coffee cans will work. Unwaxed paper cups can also be used. Most plastic cups, except those of Styrofoam plastic foam, can be used. Paper, plastic, and metal buckets available at paint stores are fine. For large laminating and lay-up jobs, small tubs are convenient for rapid handling of resin, but for most fiberglassing repair work these will not be needed. Usually only small amounts of resin are mixed (catalyzed) at a time. Short mixing containers with large bases are recommended to keep accidental spills to a minimum.

Avoid waxed containers because the wax can contaminate resin. Sometimes the resin will soften the wax and cause the bottoms to fall out of the containers.

I recommend disposable cans, cups, buckets, and tubs. After a few uses, discard them. It is much more convenient than trying to clean the containers and often less expensive, too. For polyester resin, for example, the acetone required for cleaning a container often costs more than a new disposable container. Clean mixing containers are essential if contamination of resins and other chemicals is to be avoided.

Fiberglassing workers have their own preferences for mixing containers. For example, some like buckets without handles; others prefer buckets with handles.

Have an adequate supply of mixing containers at hand before you start fiberglassing work. Because only a small amount of resin is generally mixed (catalyzed) at a time, the resin usually isn't mixed with catalyst in the container in which it is sold. Once the catalyst or curing agent is added, the resin will harden even if the lid is put back on the container. The resin is poured from the container it was purchased in into the mixing container in the desired amount, then the catalyst or curing agent is added.

Mixing Sticks

Wood, plastic, or metal mixing sticks available at paint stores are used for mixing purposes. For most fiberglassing jobs, two or more clean sticks of several lengths will be required. Small ones about the size of ice cream sticks are useful for mixing resin in small cups. Longer and bigger ones can stir resin in gallon cans before pouring out a quantity into another container for use. Once a stick has been used in catalyzed resin, it should not be used to stir uncatalyzed resin. There is likely to be enough catalyst present to set off the curing reaction.

Paint Brushes

Paint brushes can be used for applying resin to reinforcing materials. I suggest inexpensive (throwaway) brushes, as brushes are quickly ruined by fiberglassing. While solvents are available for cleaning uncured resins from brushes, these will not work on cured resins. Once the resin is allowed to harden in the brush hairs, the brush is finished. Throw it away and start again with a new one. Avoid brushes with painted handles. The resin and acetone and epoxy solvents usually cause the paint to come off.

The size of the brushes depends on their intended use. The most useful sizes for common fiberglassing repair work, I believe, are ½-, 1-, and 1½-inch-wide brushes. For large laminating and lay-up jobs, even wider brushes can be used.

Squeegees

Squeegees (Fig. 3-1) with either rubber or plastic blades are useful for scraping off excess

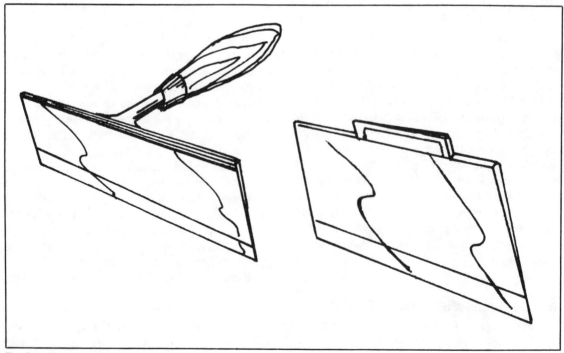

Fig. 3-1. Squeegees.

resin from reinforcing cloth and woven roving when doing laminating or lay-up work. This is important for keeping the reinforcing-material-to-resin ratio high for greater strength in the cured laminate. Squeegees can also be used for spreading resin quickly and removing air bubbles.

The blades on some squeegees are stiff. Others have some flexibility (such as those used for cleaning window glass). It will take some experimenting with both types to determine which you prefer for fiberglassing work. Squeegees are available in various widths or in strips that you can cut to desired lengths. Those with blades from about 3 to 6 inches wide are about right for most fiberglassing repair work. Some squeegees are gripped by the sides of the blade; others have handles. The choice is largely a matter of individual preference. Those with handles can be used to get at hard to reach areas.

Scissors or Snips for Cutting Reinforcing Materials

Scissors or snips are needed for cutting reinforcing materials. These should be sharp and in good working order. Ordinarily, cut reinforcing material in an area away from wet resin. If you should happen to get resin on scissors or snips, it can be cleaned off with solvent, provided that it is not allowed to cure. Use acetone for polyester resin and epoxy solvent for epoxy.

Masking Tape

Masking tape is useful for masking off areas in many types of fiberglassing repair work. It is a good idea to have several widths on hand, such as ¾, 1, and 1½ inch.

Construction Paper

Construction paper is useful for protecting and masking areas from resin. Plastic drop cloths can also be used. Make certain that they are of a plastic that will not be dissolved by the resin being used.

Protecting areas from possible resin splattering, dripping, and spilling is extremely important. Resin, especially if it is allowed to cure, is difficult to clean up from floors.

Rags

You can never get enough rags for fiberglassing work. If you don't have enough of them from old clothes and so on, you can purchase them at paint stores. They are usually fairly expensive. In some areas thrift stores sell them laundered and in bundles. You will have to remove buttons and zippers, and some items will be unsuitable for rags, but it can still be a good deal.

White rags are best, as resin and solvents often will dissolve color dyes. Some synthetics, like rayon, are not suitable. Paper towels can be used instead of rags or in addition to rags.

Putty Knives

Putty knives come with a variety of blade widths (Fig. 3-2). I suggest at least two for a start —one with a 1-inch wide blade, the other with a 2½-inch wide blade. These are useful for applying resin putties and fillers.

Cellophane

Cellophane is useful for some types of fiberglassing repair work as a release material. It seals off the air from small areas when air-inhibited resin is used so that the surface will cure tack-free.

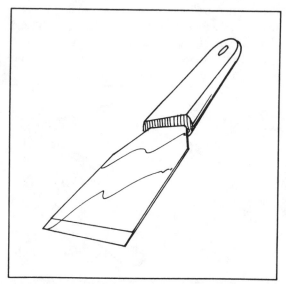

Fig. 3-2. Putty knife.

Utility or Razor Blade Knife

There are many uses for a sharp cutting knife in fiberglassing repair work. A single-edge razor blade can also be used.

Sanding Block

A small block of wood that the sandpaper can be held around, or special sanding blocks that clamp the sandpaper in place, can be used (Figs. 3-3 and 3-4). Sanding blocks are used so that sanding will not cause the surface to become uneven. This often happens when the sandpaper is held by hand using finger pressure.

Sandpaper or Abrasive paper

The main types of sandpaper or abrasive paper are garnet, Carborundum, aluminum oxide, and silicon carbide. Aluminum oxide and silicon carbide are the most satisfactory for sanding fiberglass (Fig. 3-5), though all types can be used. Sanding is generally done by beginning with coarser grits and gradually working down to finer grits. Thirty-six-grit paper is for very coarse sanding, 150-grit paper is for medium sanding, and 600-grit paper is for very fine sanding.

A selection of grits is needed for fiberglassing repair work. Coarse and medium sanding is usually done with "dry" sandpaper. Fine sanding is done dry or wet, depending on the particular job, and thus wet/dry sandpaper is recommended for fine grits. Exact selection of sandpaper will depend on

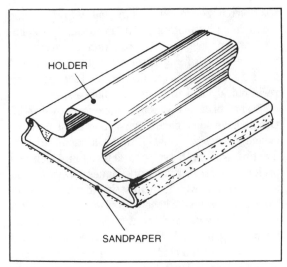

Fig. 3-4. Holder that clamps sandpaper in place for block sanding.

the material to be sanded and the particular job at hand, as detailed in later chapters.

Flat sheets of sandpaper are used for hand and block sanding. Power sanders, if used, often require special shapes and size.

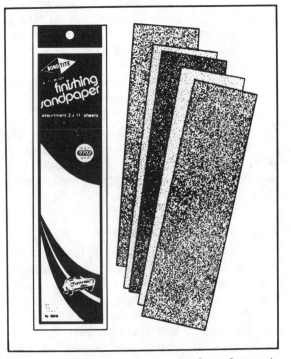

Fig. 3-5. Bond Tite sandpaper (courtesy Oatey Company).

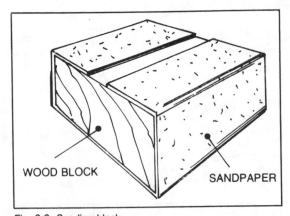

WOOD BLOCK   SANDPAPER

Fig. 3-3. Sanding block.

Fiberglass Rubbing and Polishing Compound

Fiberglass rubbing and polishing compound is required for making gel coat repairs. The compound also restores the appearance of faded and chalked gel coat surfaces.

Fiberglass Wax

Special waxes are available for use on fiberglass gel coat surfaces. These are frequently used not only for routine maintenance of fiberglass gel coat surfaces but also for fiberglassing repair work.

Surfacing Tools

Surfacing tools (Fig. 3-6) come in a variety of shapes, configurations, and sizes. These are very useful for fiberglassing repair work, especially if power grinding and sanding tools are not being used. Surfacing tools make it possible to do fiberglassing repair work, especially small jobs, without power sanders. Surfacing tools also raise less dust, making them safer to use then power sanders.

Files

Files of various shapes and sizes are useful for fiberglassing repair work. Usually metal files are used because of the abrasive nature of cured fiberglass. A typical selection for fiberglassing repair work will include a flat, half round, and round file.

Laminating or Lay-Up Rollers

Laminating or lay-up rollers (Fig. 3-7) are used to remove air bubbles from reinforcing materials that have been impregnated with wet resin. Those with aluminum rollers are the most satisfac-

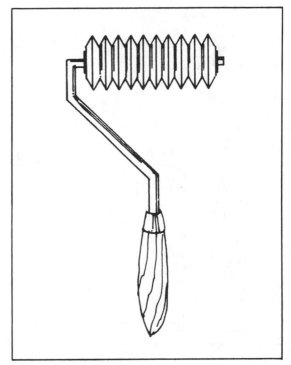

Fig. 3-7. Laminating roller.

tory. They are easily cleaned with solvent. If resin does harden on the rollers, it can be burned off with a torch. Standard rollers are typically available with 1- and 2-inch diameter rollers from about 3 to 8 inches long. Some feature spiral thread ridges to better break up air bubbles. Special corner rollers (Fig. 3-8) for working on rounded surfaces are also available.

While considerable fiberglassing repair work can be done without laminating or lay-up rollers, the rollers can be handy for jobs involving lay-up of large areas of reinforcing material. They are used extensively for fiberglass lay-up work in a mold. Some have long handles for getting at places that are difficult to reach.

Paint Rollers and Trays

Paint rollers can also be used for applying resin to reinforcing material. The catalyzed resin can be placed in plastic or metal paint trays for saturation of the roller, which is then rolled over

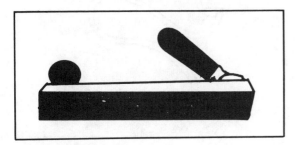

Fig. 3-6. Surfacing tool.

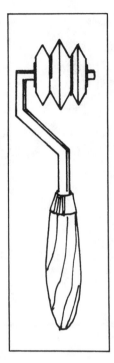

Fig. 3-8. Special corner laminating roller.

Portable Electric Drill and Attachments

While a limited amount of fiberglassing repair work can be done without power tools, they can make the work much easier and faster. If you could choose only one power tool for fiberglassing repair work, it would probably be a portable electric drill (Fig. 3-9).

A ¼-inch electric drill with variable speed can serve many purposes. With metal-cutting bits, it can be used for drilling holes in cured fiberglass laminates. With spade-type bits and hole saw attachments, even larger holes can be made (Figs. 3-10 and 3-11).

Abrasive grinding burrs, such as the one shown in Fig. 3-12, can be used in the electric drill for widening out cracks in fiberglass and for many other uses. While hand tools can be used for the same purpose, the work is generally much easier and faster with the electric drill and grinding burr attachment. Abrasive grinding burrs are available in several shapes and sizes.

A disk sanding attachment (Fig. 3-13) is useful, especially if you don't have a separate disk sander. Sandpaper in various grits is available to fit the sanding disk attachments.

The disk sanding attachments come in various diameters. The 5- and 6-inch sizes are about right for fiberglassing repair work. The sanding is usually done with the electric drill turning at a fairly high speed.

the surface being laminated for application of the resin. This method of applying resin seems inconvenient to me, but some people like it.

Paint rollers are sometimes also used for smoothing the surface of a fiberglass lay-up by placing cellophane or clear plastic film food wrapping material over the wet resin, then working the paint roller over it. This procedure can save a lot of sanding later. Sanding is difficult and time-consuming work. Manufacturers of fiberglass products try to keep the amount of sanding required to a minimum or to eliminate it entirely in some cases.

Devices for Applying External Heat

Devices for applying external heat can be useful when working with polyester resin in cold conditions. They are almost essential for working with epoxy resin if curing is to take place in a reasonable amount of time.

Infrared heat lamps, portable electric heaters, hair dryers, and heat guns can be used. Only flameless heating devices should be used.

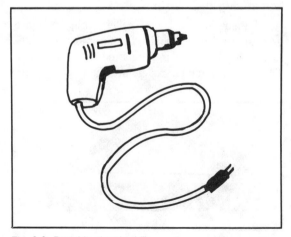

Fig. 3-9. Portable electric drill.

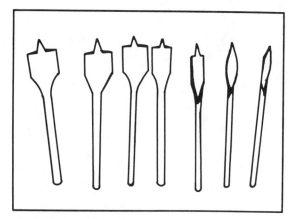

Fig. 3-10. Spade-type bits.

Polisher or buffer attachments are available to fit over the disk sanding attachments. These are useful for fiberglassing repair work. When a variable speed electric drill is used, a fairly slow turning speed is recommended. The principle is high speed for sanding, low speed for buffing and polishing.

Electric drills are mainly designed for drilling holes. They are suitable for limited sanding and buffing work, but for extensive work, more specialized sanding and polishing tools are recommended.

Because fiberglass dust is highly abrasive, it can be very hard on power tools. Frequent cleaning of tools with an air gun (attached to an air compressor) will help to protect the tools from wear and damage from this dust. When purchasing tools for fiberglassing work, select those specially designed for use in the presence of abrasive dust.

Fig. 3-11. Hole saw attachment for portable electric drill.

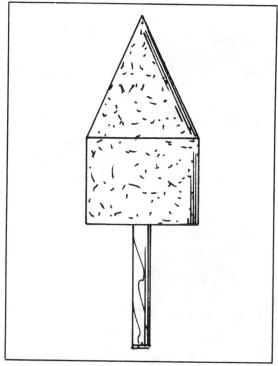

Fig. 3-12. Abrasive grinding burr attachment for a portable electric drill.

Hacksaw

Metal cutting hacksaws are useful for cutting cured fiberglass. While a regular pistol or saw grip hacksaw frame can be used, a file-type handle is often more convenient for fiberglassing repair work.

Saber Saw

Saber saws are portable power tools that can cut cured fiberglass laminates (Fig. 3-14). Metal-cutting blades should be used. Typical uses include cutting away badly damaged fiberglass in preparation for patching and making cutouts in fiberglass laminates. While the same cuts can usually be made by hand with a hacksaw, the saber saw makes this work much faster and easier.

Portable Power Sanders

Power sanders, while perhaps not absolutely essential for small jobs, are important tools for

35

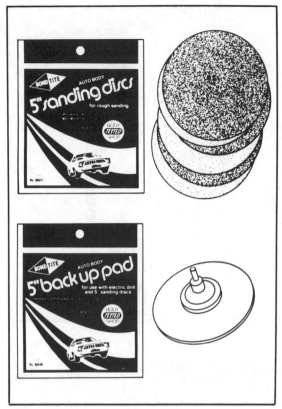

Fig. 3-13. Bond Tite sanding disk attachment for a portable electric drill (courtesy Oatey Company).

fiberglassing repair work. The three basic types are the *pad* (also called vibrating) sander, *disk* sander, and *belt* sander (Fig. 3-15).

Pad sanders are available with orbital, straight line, and combination orbital and straight line actions, with the latter being the most useful for

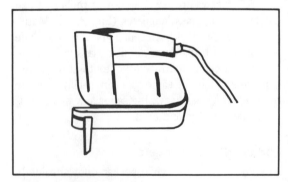

Fig. 3-14. Saber saw.

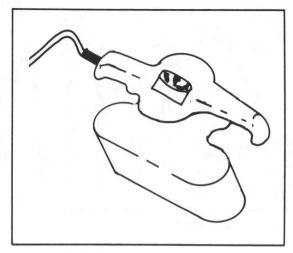

Fig. 3-15. Belt sander.

sanding fiberglass. Pad sanders are designed for finishing and light-duty sanding.

Disk sanders have disks that are usually mounted at right angles to the drive spindle. This makes sanding much easier than with a disk attachment mounted in an electric drill. Disk sanders are available with disks from about 5 to 9 inches in diameter. A heavy-duty model with a 7-inch disk will work well for fiberglassing repair work.

Some disk sanders feature a low-speed adjustment, so they can also be used with buffing and polishing pads. This tool is often called a combination disk sander and polisher. You can use a single tool for two purposes, or you can have separate tools — a high-speed disk sander and a low-speed disk polisher. The single tool is less costly; two separate tools are more convenient if a large amount of fiberglassing repair work is to be done.

Belt sanders have a belt of sandpaper that travels over two drums. These are useful for some types of fiberglass sanding work. They can be adjusted for heavy, medium, and light sanding.

Of the three types of sanders, the disk sander is generally the most versatile for fiberglassing repair work. It also takes the most skill and practice to use effectively without leaving undesirable swirls in the surface being sanded. A soft pad between the disk and the sandpaper can be helpful here. Grinding attachments can also be used with many disk sanders. These are sometimes useful for coarse work.

Pad sanders are useful for jobs requiring plenty of finishing sanding. Belt sanders can be used for light, medium, and heavy sanding when greater precision is required than is possible with a disk sander.

Anyone planning to go into the fiberglassing repair business will probably want to invest in a heavy-duty, commercial model, disk sander and take the necessary time to learn to use it properly. In skilled hands, professional sanding results can be accomplished in a minimum time period. Pad and belt sanders can also be useful, especially for certain types of sanding work.

Portable Power Polisher

This can be a combination tool with the disk sander described earlier or a separate tool. The polisher must turn at a slower speed than the sander so that heat buildup will not burn the fiberglass surface when buffing and polishing.

Small polishing and buffing jobs can be done by hand with a cloth and necessary compounds. For larger jobs, a portable power polisher can be a good investment.

Backing Materials, Wire, and Clamps

A variety of materials are needed to back up areas to be repaired. Various types of wire and clamps can hold these in place. Tools, materials, and methods will be detailed in later chapters.

Body and Fender Tools

Metalworking body and fender hammers, dollies, and other tools are useful for preliminary preparation of metal for fiberglassing repairs. Suggested tools for particular jobs and their uses are covered in Chapter 13.

Other Woodworking and Metalworking Tools

Because wood, metal, and a variety of plastics can be repaired and/or protected by fiberglassing, other woodworking and metalworking tools will be needed for some types of repair work. On fiberglass laminates, metalworking tools can often be used, as detailed in Chapter 6.

Compressor

A compressor with an air gun attachment can blow dust from power tools and have many other uses, such as for an air supply for spray guns. I do not recommend spraying resin unless you can afford to invest in very expensive protective equipment, which is beyond the scope of this book. Besides, spraying rigs designed for mixing catalyst and resin in the proper amount at the nozzle generally cost $2,000 or more. Resin can be sprayed with less expensive equipment by thinning the resin and catalyzing the resin all at once before starting the spraying. This method involves the same or even greater health hazards.

Methods for applying gel coat and painting fiberglass that don't involve spraying or can be done with spray cans are detailed in later chapters. Brushable two-part polyurethane paint is rapidly replacing gel coat as a refinishing technique for amateur application for deteriorated and damaged original gel coats. If you plan on going into the fiberglassing repair business, however, you will probably have to advance to spraying equipment and methods.

Spray Gun with Chopper

This is a device that sprays catalyzed resin with resin, catalyst, and acetone being applied under pressure to the nozzle for spraying (Fig. 3-16). The acetone is for cleaning the nozzle; there is a button for releasing it. This is combined with a chopper that cuts up fiberglass strands or rovings into short lengths and mixes them with the resin spray. This is a spray-up method for making a fiberglass molding. Laminates produced in this fashion are generally weaker than those that are hand laid-up with cloth or woven roving reinforcing material. Nevertheless, the device is widely used in fiberglass manufacturing. While this device is far too expensive a piece of equipment for most do-it-yourselfers and is not required for the fiberglassing repair work detailed in this book, it's a piece of equipment that a person going into the fiberglass-

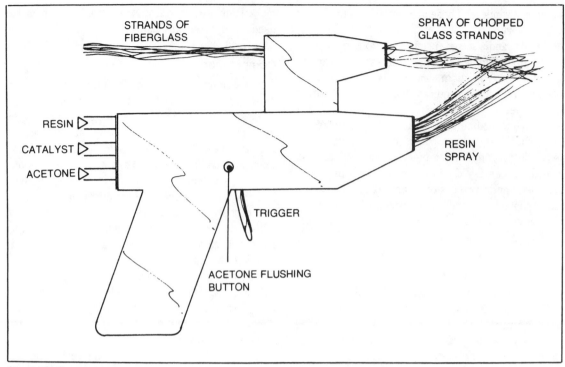

STRANDS OF FIBERGLASS

SPRAY OF CHOPPED GLASS STRANDS

RESIN ▷

CATALYST ▷

ACETONE ▷

RESIN SPRAY

TRIGGER

ACETONE FLUSHING BUTTON

Fig. 3-16. Chopper gun.

ing repair business will want to consider. The tool can greatly speed up many types of fiberglassing repair work, and the cost might not be prohibitive if it can be spread out over hundreds of repair jobs. The spray gun with chopper can be useful even for jobs that are largely hand lay-up efforts as a means of applying matlike layers between woven roving reinforcing materials. This will reduce the amount of hand labor considerably and, if done properly, will result in a laminate with adequate strength.

Protective Clothing

Protective clothing is important for health, safety, comfort, and convenience in working with fiberglass and handling, mixing, and applying the related chemicals. The protective clothing should not be thought of as a raincoat, however. Good professional house painters seldom get much paint on themselves or their clothing; beginners often get considerable paint on themselves and their clothing. The same applies to fiberglassing. When-

ever you do get resin or other fiberglassing chemicals on yourself or your clothing, you have made a mistake. The protective clothing is there just in case. The clothing protects you from air-borne dusts, mists, fumes, and chemicals.

While many types of protective clothing can be worn for fiberglassing, coveralls made from polyolefin fabric seem to work well. They are lightweight, comfortable, and easy to work in. Plus, they offer considerable protection from fiberglass dust and other particles and fiberglassing chemicals. The coveralls will protect the entire body except for head, hands, and feet.

Boots

While many types of standard work boots can be used, most will be quickly ruined by resins and other fiberglassing chemicals. It is best to set aside a pair just for fiberglassing work. Many fiberglassing workers prefer rubber rain boots, either worn alone or over shoes or boots.

You can use disposable polyethylene boots that are available from paint stores and fiberglassing material suppliers. These will fit over shoes or boots and provide good protection from resins and other fiberglassing chemicals.

Gloves

Regular plastic gloves that are resistant to fiberglassing chemicals can be used, but most do not offer enough fingertip sensitivity. Disposable polyethylene gloves do. They can be obtained from paint stores and fiberglassing material suppliers. I recommend these gloves whenever fiberglass reinforcing materials are being handled or used and whenever fiberglassing chemicals are being handled, poured, mixed, or applied.

Protective Barrier Cream

Special protective barrier cream is available for use on hands. It provides protection from polyester and epoxy resins, amine hardeners, and caustic and acidic solutions. It is a thick cream that, when applied to skin, gives protection until it wears thin or is washed off. A single application will usually give protection for two to three hours. This protective barrier cream is recommended for hands when protective gloves cannot be worn.

Hand Cleaners, Talcum Powder, and Hand Lotion

While I have observed many fiberglassing workers routinely using acetone for hand cleaning, this practice is definitely not recommended from a health and safety standpoint (see Chapter 4). Many special hand cleaners that work with or without water are on the market. They give good results for fiberglassing work without the health and safety hazards associated with using acetone for hand cleaning.

I recommend having soap or powdered hand cleaner and plenty of water handy for frequent hand washing when doing fiberglassing work. Fiberglassing workers frequently apply talcum powder and hand lotion to hands and other skin areas.

Filter Masks and Respirators

Filter or dust masks are available with replaceable filter models. These are suitable only for nontoxic dusts. The masks are frequently used when sanding fiberglass, but they do not provide protection from fumes given off by fiberglassing chemicals. Filter or dust masks must fit over the mouth and nose properly, so dust cannot get in around the edges.

A respirator gives protection against the fumes, mists, and dusts from fiberglassing chemicals and other materials. I suggest that you go to a commercial supplier of respirators for industrial use and purchase a good one. Make certain that it is approved for use with the resins and other chemicals you will be using. Get instructions on how to use it and replace cartridges. Purchase extra cartridges and other necessary supplies. I consider a respirator to be an essential piece of equipment — not optional. The respirator might be somewhat awkward and inconvenient to wear, but the protection that it provides makes it worth the trouble. I have noticed that workers in manufacturing plants tend to take the respirators off when the safety inspectors are not around, but this is just being stupid. While there might not be any apparent immediate effects from breathing the gases or vapors associated with fiberglassing work, there might well be long-term effects that will show up later. Why take the chance?

A respirator should provide the necessary protection. Secondary considerations are cost (both of the respirator and of replacement cartridges), comfortableness, and that the respirator be as light in weight as possible. Carefully follow the manufacturer's directions and instructions for using the particular respirator.

Eye, Face, and Head Protection

Eye, face, and head protection is extremely important for safe fiberglassing work. Many clear plastic eye goggles are available to provide eye protection from nontoxic dusts. Most can be worn with filter masks or respirators that do not cover the eyes.

Face shields are also available, but these are

not ordinarily used for most types of fiberglassing repair work. The shields provide protection from sparks and flying chips, but when used alone, they don't provide protection from dusts, such as those from grinding and sanding fiberglass.

Hoods are acetate visors or windows. These give protection to the head and neck and can be used while wearing some types of respirators. They cover the hair, which is important when fiberglassing overhead. Most fiberglassing repair work does not require a hard hat.

Advanced Protective and Breathing Equipment

More advanced protective systems that offer complete body protection with piped-in air are sometimes used for industrial fiberglassing work. The high cost of the equipment, not to mention the difficulties involved in operating and maintaining it, make it impractical for the do-it-yourself fiberglasser. Better, inexpensive protective and breathing equipment is also being developed.

WORK AREAS

Fiberglassing repair work should ideally be done under laboratory conditions, with controlled temperature, humidity, ventilation, and so on. Most fiberglassing repair work must be done in work areas that are less than ideal.

Considerable fiberglassing repair work must be done "on location." For example, it is impractical to move swimming pools, shower stalls, and other large and/or installed fiberglass structures to a fiberglassing shop for repair work. Instead, they are repaired where they are located. You take the required tools, equipment, materials, and supplies to the job and do the work there.

Fiberglassing work can be done inside garages and other buildings if there is adequate ventilation, or fans drawing air out of the building can be installed. Fiberglassing work can create a mess, so take this into consideration when selecting work areas.

Some fiberglassing repair work can be done outdoors in the open or, better yet, under a patio roof. Auto body fiberglassing repair and customizing and repair of small boats are often done in these work areas. Working in direct sunlight can have adverse effects on fiberglassing, so it's best to have some type of roof or shield to block the sun.

Some fiberglassing repair work can be done on boats in the water; in other cases the boats will have to be hauled out on a trailer or at a boat yard. Again, this might require working under less than ideal conditions, but adequate repair can usually be accomplished with reasonable care and planning.

Sometimes special structures are built especially as work areas for doing fiberglassing repair. This usually isn't practical unless you plan to do considerable work or are going into the fiberglass repair business.

Chapter 4

Health and Safety

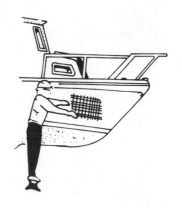

Health and safety in fiberglassing repair work are extremely important considerations. This is true of wood, metal, concrete, and other types of repair work. Each presents certain health and safety hazards and has its own rules of safety. Fiberglassing is quite different from most types of repair work. It is often more like working in a chemical laboratory than in a typical shop. There are actually two main aspects to fiberglassing: the chemical part of working with liquid resins — adding catalysts or curing agents, mixing them together, applying this mixture to reinforcing material, and so on; and the shopwork part of working with cured fiberglass — sanding it, sawing it, drilling holes in it, bolting two separate fiberglass moldings together, and so on.

Remember that in the first aspect you are dealing with fiberglassing chemicals; in the second aspect you are working with cured fiberglass — things that are quite different from wood, metal, and concrete. While there is some overlapping, there are particular tools, equipment, and materials that are being worked with.

Fiberglassing repair work is reasonably safe for the average person to undertake if certain health and safety precautions are followed. Anyone who has any reason to believe that he or she may react badly to the chemicals and materials involved should check with a physician before undertaking fiberglassing repair work.

Anyone who undertakes fiberglassing repair work must learn the rules of safety and establish and practice safe working habits. Most of the materials used in fiberglassing repair work are reasonably safe, but they must be treated with considerable respect. Unfortunately, I've seen many instances where this attitude has not prevailed. Some fiberglass workers violate safety rules to prove how tough, brave, and manly they are. Typical examples include not using respirators, smoking while using flammable and even explosive chemicals, and not wearing protective clothing.

SAFETY RULES

The basic assumptions are that health and safety are considered important and that safe working habits are worthwhile.

Ventilation

There must be adequate ventilation in the area where you are working. Prolonged and repeated exposures to high concentrations of fumes from resins and other fiberglassing chemicals can result in many health problems. There can also be immediate problems, such as skin and eye irritations.

While protective clothing and equipment can reduce these hazards, they should always be in addition to, not in substitution of, adequate ventilation. When working outdoors in the open or under cover, ventilation is rarely a problem. When there is a breeze or wind, always try to work between the direction the wind is coming from and the area where resins and other chemicals are being used. This way fumes are blown away from you and not toward you.

Inside work areas should have large areas open to the outside, preferably on at least two opposite sides, or a ventilation system. The system should have a fresh air supply and a large exhaust fan capable of changing the room air completely at least once every five minutes, but more often is much better. The exhaust fan must direct the air safely away from the building, not toward a house or other areas where people and other living things are likely to be.

Jobs that must be done on location present special problems. To repair a shower stall in a bathroom, place a portable exhaust fan in a window. To do fiberglassing work inside the cabin of a small boat, rig the exhaust fan up to a hatch. In addition, you should have as many openings as possible for maximum fresh air supply. The air inside must be changed with fresh air as rapidly as possible, within limits.

Protective Clothing and Equipment

Wear protective clothing and equipment when handling and using fiberglass reinforcing materials. Fine glass threads and particles from glass fiber reinforcing materials can be irritating to the skin, respiratory system, and eyes. Exposure can result in allergic reactions, skin rash, dermatitis, and other problems. A first step in preventing this is to minimize or eliminate contact between the glass fiber material and the body. Wear the protective clothing and use the protective equipment described in Chapter 3. Do not handle fiberglass reinforcing materials with bare hands. Wear protective gloves, such as the disposable polyethylene type. This might seem like a needless precaution when, as often happens, the clerk at the store where the material is purchased handles it with his or her bare hands. This careless handling of the material probably adds not only oils to the fiberglass reinforcing material, which makes it more difficult to saturate with resin, but also gives the clerk itchy skin or worse.

The sensitivity to glass fiber reinforcing mate-rial varies greatly from individual to individual. Some people handle it constantly without protective equipment with bare hands. They have little or no discomfort or apparent health problems. There might well be long-range problems, however.

A few people have problems with glass fiber reinforcing material even when they wear the proper protective clothing and equipment. For those with this extreme sensitivity, there are several possible solutions. The first is to quit fiberglassing work. The second is to seek advice from a physician. The third is to use one of the nonfiberglass reinforcing materials such as polyester, polypropylene, or acrylic instead of fiberglass. These nonfiberglass reinforcing materials generally do not cause skin irritation and allergic reactions.

If fiberglass reinforcing materials are used, store them in plastic bags to reduce the amount of glass fiber strands and particles that get into the air. This is especially important in the case of fiberglass mat, where small pieces of glass fiber come off fairly easily.

Take special care when cutting fiberglass reinforcing material, as this also tends to release glass fiber into the air. After cutting, use a vacuum cleaner to pick up any loose strands and particles of glass fiber from the cutting area. The cutting area can be a floor or table. It should be clean and, if desired, covered with plastic or paper, which can be held in place with masking tape. Keep resins and other chemicals away from the cutting area.

Some people don't like to use scissors or shears while wearing gloves. With a little practice, scissors or shears can easily be operated while wearing the thin disposable polyethylene gloves, which give good fingertip sensitivity.

If, despite all precautions, glass fibers do get on the skin, a cold shower followed by application of hand lotion often helps relieve itching. If skin rash or any other unusual reactions develop, consult a physician.

Wear protective clothing and equipment when sanding fiberglass. Sanding cured fiberglass produces a fine dust that can cause respiratory, eye, and skin problems. Wearing proper protective clothing and equipment (see Chapter 3) while sanding fiberglass is extremely important. A well-fitted filter mask or respirator and eye goggles or protective hood are especially important. Do not overlook protecting the skin, including your hands.

Even hand sanding can raise considerable dust. Hand sanding with or without a block can usually be done while wearing polyethylene disposable gloves.

Power sanding presents even greater dust problems. Even bearings and brushes in power tools tend to suffer the effects, so the importance of not breathing the dust or getting it in your eyes or even on your skin is rather obvious. Fully body protection is extremely important. One problem is operating power sanders while wearing gloves. With a little practice, this can be done, but fairly substantial gloves are generally required. The thin disposable polyethylene type wear through quickly. Rubber and plastic gloves with liners will work satisfactorily. Chemical resistance is not a critical factor here as the sanding is done on cured fiberglass.

When sanding laminates where nonfiberglass reinforcing materials were used, the effects of dust contact with skin are usually less severe than when glass fiber reinforcing materials are used in the laminate.

Sanding dust that is allowed to accumulate in the work area can also cause problems as some of it can easily become airborne. Frequent cleaning of the work area with a shop-type vacuum cleaner will help to alleviate this problem.

If any respiratory, eye, or skin problems develop, consult a physician. This should rarely be necessary if you observe the safety precautions for sanding fiberglass, and use proper protective clothing and equipment.

If, despite all precautions, some sanding dust does get on the skin, this usually isn't much of a problem. A cold shower followed by application of hand lotion will often help to relieve itching. If skin rash or any other unusual reactions develop, consult a physician.

Wearing proper protective clothing and equipment for fiberglass sanding might be uncomfortable, especially in hot weather. I believe that this slight discomfort and inconvenience is a necessary sacrifice for the protection of your health.

Follow Manufacturer's Recomendations

Follow manufacturer's recommendations for safe use of products. Always read the labels on fiberglassing resins and other products not only to learn how to use them, but also to learn what safety precautions to follow. Follow directions carefully and observe all health and safety precautions. Sometimes there will be warnings, such as: "May be harmful or fatal if taken internally. Avoid prolonged exposure to fumes. Use only in well-ventilated areas. Avoid skin contact."

Often the steps to be taken in case of an accident, such as splashing resin in the eyes, are given. Always know what these are before you use the product, just in case.

Keep Fiberglassing Chemicals Away from Fire

Keep resins, catalysts, curing agents, solvents, and other fiberglassing chemicals away from fire and flame. Most polyester resins are flammable in both liquid and cured states. Even fire-retardant polyester resin is flammable when in liquid state. The same applies to epoxy resins and epoxy curing agents or hardeners, with the added problem that some will explode. Appropriate warnings should be on containers' labels. Because spontaneous combustion is a possibility, store these chemicals properly. Heat from fire or flame can cause catalysts for polyester resins to explode. Acetone and epoxy solvents are both highly flammable. They give off heavy vapors that tend to travel along close to the floor. If the solvents reach fire or flame, they will explode. A spark can cause the same thing to happen. Handle these solvents with the same care and caution as gasoline.

Avoid Open-Flame Heaters

Do not use open-flame heaters in fiberglassing work areas. This is an extremely important consideration. If the work area is to be heated, make certain that the heater used does not create a fire and explosion hazard with the chemicals and materials being used. The same applies to heating devices used for speeding cure of resins.

When selecting and installing heaters for use

in fiberglassing work areas, get professional advice. When doing fiberglassing work on location, make certain that there are no open-flame heaters, stoves, or other devices in the area that could create a fire or explosion danger with the chemicals you will be using.

No Smoking

Do not smoke in fiberglassing work areas, especially when resins and other chemicals are being used. This might seem obvious, but I think it should be mentioned. Some people don't take the obvious seriously. If you must smoke, take a break from the fiberglassing work and do it away from the work area, especially away from the resins, solvents, and other chemicals.

Storage of Fiberglassing Chemicals

Store fiberglassing chemicals in cool, dry places. This will not only increase the shelf life of resins and some of the other fiberglassing chemicals, but it will also help to prevent the possibility of spontaneous combustion or explosion. Never store the chemicals in the sun or near fire or flame.

Avoid Skin Contact with Fiberglassing Chemicals

Avoid skin contact with resins, catalysts, curing agents, solvents, and other fiberglassing chemicals. Wear protective clothing and use protective equipment to keep the fiberglassing chemicals from getting on your skin. Take care when handling, pouring, mixing, and applying the chemicals.

Some of the chemicals are potentially more dangerous than others. Be especially careful of methyl-ethyl-ketone (MEK) peroxide catalyst for polyester resins and epoxy resins and curing agents, which can produce chemical burns if they come into direct contact with the skin. Even polyester resin, generally considered to be much safer to use than epoxy, can cause skin irritations, dermatitis, and other health problems from skin contact.

No Eye Contact with Fiberglassing Chemicals

Avoid eye contact with resins, catalysts, curing agents, solvents, and other fiberglassing chemicals. The consequences of splattering or otherwise getting even small quantities of these chemicals in the eyes can be extremely serious. This must be avoided. The first line of prevention is careful handling, pouring, mixing, and application of the chemicals. The second line of defense is to always wear goggles when using these chemicals. Finally, always know the emergency steps to take it any of the chemicals being used do contact the eyes. This varies depending on the particular chemical. Sometimes the eyes must be flushed with large quantities of water, and Emergency medical treatment is needed as soon as possible. The specific instructions should be on the label of the product's container.

Avoid Breathing Fumes from Fiberglassing Chemicals

Avoid breathing fumes from resins, catalysts, curing agents, solvents, and other fiberglassing chemicals. This applies to individual chemicals after they have been mixed, such as catalyst or curing agent added to resin. Work only in well-ventilated areas. Wear a well-fitted respirator that is safety approved for the particular chemicals you are using. For the respirator to remain effective change the cartridges or elements as directed by the respirator's manufacturer. In addition, keep chemicals as far away from your face as possible. This is an extremely important safety rule—one that too often isn't given the attention it deserves. While some breathing of fumes from resins and other fiberglassing chemicals is almost inevitable in fiberglassing repair work, keep it to an absolute minimum.

Other Rules

☐ Never mix a polyester accelerator like cobalt napthenate directly with a catalyst such as methyl-ethyl-ketone (MEK) peroxide. An explosion can occur. Most manufactured polyester resins already have the necessary accelerator mixed in at the factory. If you do add an accelerator, add it to and mix it with uncatalyzed resin *before* adding the catalyst.

□ *Methyl-ethyl-ketone (MEK) peroxide should not be stored in metal containers as spontaneous explosions can occur as a result of prolonged contact with metal.* This catalyst usually comes in plastic containers. Do not put this catalyst in metal containers for storage.

□ *If you do get resins and other fiberglassing chemicals on your skin, remove them as soon as possible.* Use a rag to wipe them off. Use soap or hand cleaner and water to wash them off—the sooner, the better. Problems from skin contact with chemicals depend, among other things, on the particular chemical, the amount, and the length of exposure.

□ *Develop clean working habits.* Keep the work area clean and organized. Discard floor and bench covers, disposable containers, and similar items when they are no longer serviceable. Clean up spilled chemicals. Clean working habits can make fiberglassing repair work safer and more enjoyable.

□ *Keep visitors away from the work area.* Don't allow other people without protective clothing and equipment to be in the work area.

□ *Keep children away from the fiberglassing work area.*

□ *Keep pets away from the fiberglassing work area.*

□ *Have the best working conditions possible.* Good lighting, comfortable working temperatures, and other related factors all help to make fiberglassing repair work safer and more enjoyable.

□ *Follow all safety precautions and use proper safety equipment when using power tools.* Always follow safe operating procedures for portable electric drills, saber saws, sanders, and other tools used for fiberglassing repair work. This is the non-chemical part of fiberglassing repair work. Don't get careless.

□ *Have one or more fire extinguishers handy.* Always have one or more approved fire extinguishers available for immediate use in the work area in case of a fire emergency. These should be of a type that will work on chemical fires.

□ *Keep a first aid kit handy.* This kit is a good idea for almost any type of repair work. Minor cuts, blisters, and so on should be treated to prevent possible infection.

□ *If you work alone, have assistance nearby.*

DEVELOPING SAFE WORKING HABITS

Knowing the rules of safety is one thing; practicing them is quite another. A do-it-yourselfer can usually follow or disobey safety rules as he or she pleases. A fiberglassing worker employed in a manufacturing plant is not given this choice. Using safety equipment and following safety rules, including those that apply to the worker's own health, are part of the job, at least when the safety inspector is around. Allowing workers to violate safety rules and precautions can result in heavy fines.

The prudent do-it-yourself fiberglasser develops safe working habits because he or she wants to. The basic idea is to follow the rules of safety right from the start. In this way they will soon become habit. Safety rules will become a routine part of fiberglassing, and you won't have to consciously think about them. The breathing respirator and gloves automatically go on for a job that calls for them, and so on.

On the other hand, not following safety rules and precautions can become habit, too. You will only be working with the chemicals for a few minutes, so you don't bother with the respirator. Nothing much seems to happen. Next time you do an even longer job without the respirator. This practice continues until working *without* the respirator becomes habit. This is to be avoided.

There are many reasons why fiberglass workers violate safety rules. Respirators tend to be cumbersome to wear. If you are going to do fiberglassing work, though, they are worth getting used to.

Violation of safety rules often results in no apparent ill effect on the person who violates them. Even if there is no immediate effect, there might well be a long-term effect.

There is the attitude that it "won't happen to me." Maybe not, but why take the chance?

Finally, don't let the requirements for safe working in fiberglassing scare you out of doing your own fiberglassing repair work. The safety precautions that should be taken are not excessive in comparison to many things that you commonly do (for instance, operating a car). They are just less familiar. Fiberglassing repair work can be a safe and rewarding experience.

Chapter 5

Care and Maintenance of Fiberglass

An important advantage of using fiberglass as a construction material is that it is an easy material to care for and maintain. Fiberglass can suffer damage by deterioration, wear, abuse, and accident. This can be cosmetic, structural, or a combination of the two.

The amount of care and maintenance necessary to keep fiberglass surfaces looking new and prevent or slow deterioration depends on the particular construction and how and where it is used, among other things. Fiberglass boats and car bodies must be protected from the effects of the sun's rays; bathtubs and shower stalls located indoors don't have to be.

The surface of fiberglass on one or both sides can be finished with gel coat, which is usually molded in as part of the laminate, or by painting. In either case, the care and maintenance is basically the same.

Cleaning, polishing, and waxing fiberglass surfaces, whether gel coated or painted, keeps them looking new and slows deterioration. If the surfaces have faded or chalked from lack of maintenance or other reasons, cleaning, polishing, and waxing will often give new life to these surfaces. Consider these steps before taking more drastic repair measures, such as applying new gel coat or painting the surfaces (see Chapter 10).

CLEANING

Household detergents or special fiberglass soaps can be used for washing and cleaning most dirt, stains, and oils from fiberglass surfaces. For more difficult cleaning, many products are available especially for fiberglass. Always use the mildest cleaning detergent or solvent that will get the job done.

Stubborn stains, tar, and grease might require the use of scouring powders, abrasive pads, or polishing compounds. Because these tend to wear some of the gel coat or painted surface away, use them with discretion. Combination cleaners and polishes specially formulated for fiberglass surfaces, such as those available in spray cans, will do the job more safely. When abrasives are used, polishing will be necessary to restore luster to the surface.

POLISHING AND WAXING

After cleaning with detergent and water, apply a liquid or paste cleaning and polishing compound to the surface. This is an infrequent necessity, perhaps only once or twice a year. For gel coat surfaces that have faded, chalked, or suffered considerable oxidation, use a special fiberglass rubbing compounds, which is readily available.

After cleaning and polishing, apply a good grade of fiberglass wax if desired. There are differences between brands, and in some cases little or no distinction seems to be made between the terms "polish" and "wax." Some seem to do a better job than others but are perhaps difficult to apply or don't last long. Others are easy to apply. Some are simply added to the wash water but perhaps don't

47

do a very good job. Some require considerable hand rubbing or "elbow grease."

Experiment to find out what works best for your particular purposes. While polishing and waxing can be a lot of work, it only has to be done once or twice a year for most fiberglass products.

A power polisher with a buffing pad can also be used. First, thoroughly clean the surface. Next, apply buffing compound and use the power polisher. Hold it flat on the surface and apply only light pressure. Keep it moving. Use caution to avoid excessive wear on the gel coat or painted surface. Follow with a coat of wax, which can also be buffed out with the power polisher.

Many fiberglass cleaning, polishing, and waxing compounds are now on the market, as well as combination products that are formulated to do two or more of the previously mentioned jobs. I have had good results with a waxless fiberglass silicone boat polish. According to the label, the product contains no abrasive chemicals.

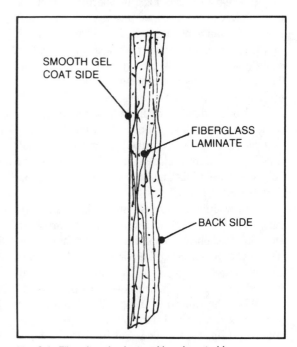

Chapter 6

Working with Cured Fiberglass

Before attempting the chemical part of fiberglassing, it's a good idea to become familiar with cured fiberglass laminate. You should learn how to saw, file, sand, and drill holes in it, and how to use prefabricated fiberglass laminate as a construction material.

CURED FIBERGLASS LAMINATES

If possible, obtain scrap pieces of cured fiberglass laminates, preferably with a color gel coat side on them. These can be used for examining the material and learning how to use tools on cured fiberglass.

Figure 6-1 shows a typical fiberglass laminate. The smooth color side is the gel coat, which was applied first against a mold (this is just one of a number of fiberglass molding methods; see Chapter 8). Behind this is a hardened mixture of plastic resin and reinforcing material. The other side of the laminate can be rough or smooth, depending on the molding and laminating method that was used (see Chapter 8).

Using an awl or other sharp-pointed metal object, try to scratch and chip first the gel coat side of a scrap fiberglass laminate, then one edge, and then the back side. While the ease with which this can be done will vary depending on the makeup of the particular piece of fiberglass, the gel coat generally forms a hard protective layer. The gel coat chips rather easily, as it is a thin layer of resin without reinforcing material. The edge of the lami-

nate can often be chipped rather easily because of delamination between layers of reinforcing material (assuming hand lay-up with several layers of reinforcing material is being used). Resin alone tends to be brittle. It will chip rather easily. Combined with reinforcing material, it is stronger and less easily chipped. The more reinforcing material in comparison to the amount of plastic resin, the stronger the laminate tends to be, provided that

Fig. 6-1. Fiberglass laminate with gel coat side.

the reinforcing material is fully saturated with plastic resin.

Place a small piece of cured fiberglass in a vise and try to break it by hitting it with a hammer. Does it snap in two? Does the resin break and the reinforcing material remain intact? The answer depends on the particular laminate and the way it was struck, but a fiberglass laminate will not crack cleanly into separate pieces as a piece of ice will. Because of the reinforcing material in the laminate, the cured fiberglass tends to break or shatter as a unit.

USING TOOLS ON CURED FIBERGLASS

In the course of fiberglass repair work, there are many situations where it will be necessary to work with cured fiberglass laminates, both original laminates and those laid up in the course of repair work. Frequent operations included drilling, sawing, filing, and sanding. Fiberglass has different characteristics than wood or metal.

Drilling

It is easy to drill holes in fiberglass. While a hand drill can be used, a portable electric drill makes the work much easier. Metal twist drills or bits are commonly used for making small holes. For slightly larger holes, spade-type bits can be used. These are convenient for holes from about ⅜ inch up to about 2 inches. You must have a different bit for each desired hole size. Even larger holes can be made with hole saws. These often come in sets with blades for holes from about ¾ inch up to about 3 inches. A different blade is used for each desired hole size. Spade bits and hole saws can be used in portable electric drills.

Before drilling, carefully mark the desired location for the hole. Use a center punch or other sharp-pointed object to make a small indentation or pilot mark for centering the point of the drill or bit. Do this carefully so as not to chip the gel coat beyond the area where the hole is to be located.

While the laminate can be drilled from either side, it's best to drill with the bit starting on the gel coat side (Fig. 6-2). If you drill from the reverse side, there is considerable danger of chipping the

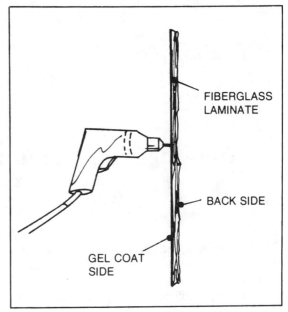

Fig. 6-2. Drilling fiberglass.

gel coat beyond the hole area when the twist drill, spade bit, or hole saw blade goes through the laminate.

Center the point of the drill or bit in the small indentation or pilot mark. Angle the drill as desired and drill a hole through the fiberglass. When using a hole saw, if you can get to both sides of the laminate, drill halfway through. Finish the drilling from the opposite side using the same pilot bit hole.

For larger holes, if you require a hole size between two sizes of spade bits or hole saws, drill the hole one size smaller. File the hole out to the desired size. Filing is covered later in this chapter.

Using scrap pieces of fiberglass laminate, practice drilling various size holes. Try to make clean holes with no chipping of resin around the hole.

Sawing

There are many situations in fiberglassing repair work where sawing of cured fiberglass laminates is required. The sawing can be by hand with a hacksaw or coping saw with a fine-tooth blade, or you can use a portable power saber saw with a fine-tooth metal-cutting blade.

50

Whenever possible, do your sawing from the gel coat side (Fig. 6-3). This will help to prevent chipping of the gel coat. It is sometimes helpful to place masking tape over the fiberglass on the gel coat side in the area where the cut is to be made. Sawing lines can be marked on the masking tape with a pencil. The pattern should be carefully marked. When making crucial cuts, leave a little extra and then use a file to take it to final size.

Cutouts and holes larger than available drill and hole saw sizes are usually made by sawing. First, carefully mark the pattern. Drill a pilot hole for starting the saber saw blade. Saw around the pattern line. When completing the cutout, hold the cutout piece of fiberglass so that it will not break off. This could cause chipping of the gel coat. If the cutout or hole must be an exact size, make it slightly smaller and then file to final size.

Practice making both straight and curved cuts on scrap pieces of fiberglass. Small pieces can be held for cutting by clamping them between two pieces of wood in a vise.

Filing

Metal files can be used for filing fiberglass. The filing is generally done on edges of laminates. Whenever possible, position yourself on the gel coat side of the fiberglass laminate (Fig. 6-4). File from side to side or apply pressure on the forward stroke only, lifting the file off the work when drawing it back. These methods will help prevent chipping the gel coat. Edges can be filed flat or

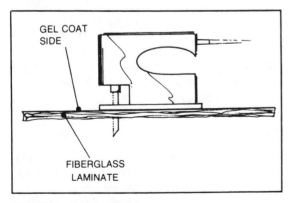

Fig. 6-3. Sawing fiberglass.

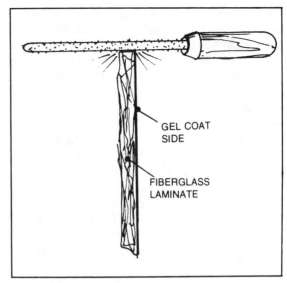

Fig. 6-4. Filing a fiberglass edge.

rounded. If the file notches become clogged with fiberglass, use a file brush or solvent to clean the file notches. Practice filing on scrap pieces of fiberglass laminate.

Surfacing Tools

Surfacing tools can be used for coarse work on edges and surfaces of fiberglass laminates, especially if power sanders are not used. They are useful for surfacing work after a fiberglass repair laminate has been laid up and allowed to cure. They can remove a large amount of material quickly.

The blades tend to clog and dull rather quickly, so frequent cleaning of the blades with solvent and replacement of dull blades is necessary. The blades on surfacing tools cannot ordinarily be resharpened; they must be replaced when they become dull or break.

Because surfacing tools tend to raise up dust, always wear a filter mask or respirator and eye protection. Keep as much of your body covered as possible when using surfacing tools.

Hand and Block Sanding

Observe safety precautions given in Chapter 4 for sanding fiberglass. Wear a filter mask or respirator, eye protection, and protective clothing.

Use aluminum oxide or silicon carbide sandpaper or abrasive paper for sanding fiberglass. Start with the coarsest grit required for the particular job and then progressively work down to finer grits. Do coarse and medium sanding dry. Fine sanding is often done wet using a sandpaper with a waterproof adhesive. Dip it in water to keep the grit from clogging while sanding and to keep scratching of the surface to a minimum.

Coarser grits of sandpaper remove the most material but also leave the deepest scratches. Medium grits remove these scratches and leave smaller scratches. Finer grits leave even smaller scratches, which are then filled in when the surface is painted or polished. Care must be taken when using coarse grits of sandpaper to avoid making scratches that extend below the desired finished surfaces. Otherwise, these will have to be filled in with fiberglass putty or other surfacing compounds to achieve a smooth finish.

Fold the sandpaper and hold it between the thumb and fingers for very fine sanding. With coarser sandpaper, however, this method can result in an uneven wavy surface. Most hand sanding is done with a block or flexible sandpaper holder with a plane-type handle. Block sanding lets you remove high spots without affecting adjacent low areas.

Disk Sanding

While small fiberglassing repair jobs can be accomplished by hand sanding, power sanders are almost indispensable for large repair jobs. A disk sander is the most useful all-around type for fiberglassing repair work. These usually have turning speeds of about 5,000 revolutions per minute. A large amount of material can be removed with a minimum amount of effort.

It takes practice to learn how to use a disk sander. When improperly used, the surface can be gouged or too much material can be sanded away. To smooth out uneven surfaces, hold the sanding disk nearly flat to the surface, apply light pressure, and keep the disk moving.

The disk sander is also useful for feathering edges. To accomplish this, hold the sanding disk a slight angle to the surface.

Anyone who plans to do a lot of fiberglassing repair work should take the time and trouble to learn to use a disk sander properly. This is especially true if you plan to go into business making fiberglassing repairs.

Belt Sanding

Belt sanders are useful for some types of fiberglassing repair work, especially when large flat areas require sanding. Like disk sanders, they can be used to remove a lot of material fast and to smooth out uneven surfaces. They can also be adjusted for fine sanding.

It takes practice to learn how to properly use a belt sander. When improperly used, the surface can be gouged or too much material can be sanded away. This is especially true on curved surfaces, where belt sanders are generally not recommended.

Some fiberglass repair shops make considerable use of belt sanding. Others don't use it at all.

Pad or Finishing Sander Use

Pad or finishing sanders are frequently called "vibrator" sanders. These sanders won't take much material off, so they are only suitable for finishing sanding and other light sanding tasks. The sander is the easiest for a beginner to use, however, as gouging the surface is seldom a problem. Pad sanders can be useful for fiberglassing repair work requiring a large amount of finishing sanding.

These sanders are available with straightline (back and forth), orbital, and combination actions. While all three types can be used on fiberglass, the orbital and combination types are preferred.

When using pad sanders, apply only light pressure and keep the pad moving. Excessive pressure can cause the shaft to break off on the sander, especially on light-duty types.

Other Tools and Jobs

Grinding tools, such as an abrasive grinding burr used in an electric drill, can be used on fiberglass. They can widen cracks in a laminate so that filler can be applied. Power buffers can be used to polish fiberglass gel coat, as detailed in Chapter 5. These buffers should turn at about 2,500 revolutions per minute, about half the speed of a typical disk sander, to prevent the pad from heating up to the point where it can burn the fiberglass. Higher speed causes more heat.

Fiberglass cannot be bent by hammering or other means. Hammering will only shatter the resin in the laminate. Even heating will not soften the laminate so that it can be bent, although enough heat will cause the resin in the laminate to burn. Fiberglass does have a certain amount of flexibility, but this allows only slight bending. Generally, fiberglass is molded to the desired shape and not bent or forced into any other shape after it has cured.

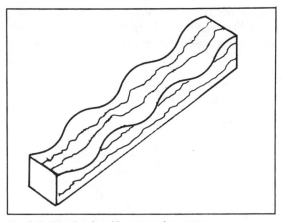

Fig. 6-5. Wood strip with corrugation pattern.

USING PREFORMED FIBERGLASS LAMINATES AS BUILDING MATERIALS

Manufactured flat and corrugated fiberglass panels are available for construction purposes. These are available in various colors, both translucent and opaque. Typical construction uses include patio roofs, greenhouses, room dividers, and decorative panels.

Many roofing accessories are available for use with corrugated fiberglass panels, including aluminum nails with neoprene washers, caulking, sealer, wood strips with a corrugation pattern (Fig. 6-5) and various types of aluminum flashings. The corrugated panels are available with regular 2½-inch corrugations 9/16 inch high (Fig. 6-6) and 4-inch ribs ⅝ inch high (Fig. 6-7). Typical panels are available in 26-inch widths in 96, 120, and 144-inch lengths. Flat panels are commonly available in the same lengths in 24, 36, and 48-inch widths.

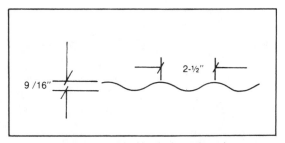

Fig. 6-6. Fiberglass panel with regular corrugations.

The panels can be sawed, drilled, and filed as detailed earlier in this chapter. Do not try to drive nails through the fiberglass panels; drill a hole. Use neoprene washers and/or caulking when installing nails to prevent leaking.

Most building suppliers who sell fiberglass panels have free building plans for patios and other constructions. To date, only limited use of the con-

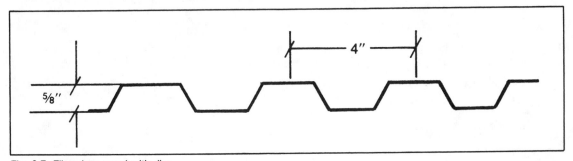

Fig. 6-7. Fiberglass panel with ribs.

cept of preformed fiberglass as a construction material has been made. If thicker, higher quality flat sheets were available, many constructions would be possible by using fiberglass bonding strips (reinforcing material and catalyzed resin) for joining the pieces together.

Chapter 7

Fundamentals of Fiberglassing

The main concern of this chapter is the chemical part of fiberglassing. No prior experience in fiberglassing is assumed. Actual practice fiberglassing exercises are detailed. These exercises are designed for learning the characteristics and applications of both polyester and epoxy resin to reinforcing materials. You will get the experience in fiberglassing necessary for doing fiberglassing repair work and molding, as detailed in later chapters.

You should be familiar with and follow the safety rules and precautions given in Chapter 4. In addition, read the manufacturer's directions and safety precautions for each product and follow them carefully.

To gain a better understanding of fiberglassing materials and techniques, both polyester and epoxy resins will be used with many reinforcing materials. These will be used for practicing, learning, and experimenting rather than actually repairing anything. I feel that this is the best approach to learning the skills and techniques necessary for fiberglassing repair work. It is better to use up some materials to gain necessary skill, control, and confidence than to begin a fiberglassing repair job immediately and botch it.

Let's begin with polyester resins. These are less expensive and easier to use than epoxy resins.

CATALYZING POLYESTER RESINS

The curing process for polyester resins is called *polymerization*. It is fully initiated when a catalyst, usually methyl-ethyl-ketone (MEK) peroxide, is added. The amount of catalyst to be added depends mainly on the working temperature, the amount of resin to be catalyzed, and the working life desired. With a typical polyester resin, ½ percent of catalyst by volume will give a working life or gel time (time from when the catalyst is added until it achieves a gelatinous consistency) of about 60 minutes at 75 degrees Fahrenheit. The working life or gel time is also called *pot life*. It's the time you have for applying the resin. After the resin has started to gel, the remainder in the pot or container should be discarded. If applied to reinforcing material, it will result in a lumpy mess. The resin usually becomes hard soon after this, although the complete cure goes on for two weeks or more. Catalyzed polyester resin will actually cure faster in the container in which it is mixed than when spread thin, such as when applied to reinforcing material.

Remember that different brands and types of polyester resins will not all cure at the same rate when the same amount of catalyst is added. Follow the catalyzing directions for the particular polyester resin being used. This can vary somewhat from the amounts given as samples in this chapter.

In the previous example, if the amount of catalyst was doubled to 1 percent of catalyst by volume, the pot life would be reduced to about 30 minutes at the same 75-degree Fahrenheit working temperature. Two percent of catalyst by volume would further reduce the pot life to about 15 minutes.

If the working temperature in these examples is 90 degrees Fahrenheit and the same amounts of catalyst are added, the pot life in each case would be approximately cut in half. It would reduce to about 30 minutes with ½ percent of catalyst by volume added, to about 15 minutes with 1 percent of catalyst by volume, and to about 7½ minutes with 2 percent of catalyst by volume.

If the working temperature in these examples is 60 degrees Fahrenheit and the same amounts of catalyst are added, the pot life in each case would be approximately doubled. It would increase to about 2 hours with ½ percent of catalyst, to about 1 hour with 1 percent, and to about 30 minutes with 2 percent.

For most fiberglassing repair work, small quantities of resin, usually from about 2 to 8 ounces (½ pint), will be catalyzed for use at a time. If larger quantities are used, there is danger of the pot life expiring before all of the resin can be applied. Once the resin starts to gel, it should be discarded. Because high heat buildup can occur with resin in a container, it is best to pour the resin out in a safe place on the ground. Allow it to harden before discarding it in a trash barrel.

As experience is gained, you might be able to catalyze and use larger amounts of resin at a time for some types of fiberglassing work, perhaps a quart or more at a time. For learning purposes, start with small amounts and then gradually work up to larger amounts as your skill and confidence improve.

Adding Catalyst

You must decide how to add such a small quantity of catalyst. For most work, the amount of catalyst added will range from about ½ to 4 percent by volume. Often the catalyst is supplied in dispensers that allow release of the catalyst by drops. In order to catalyze 1 ounce of resin with 1 percent by volume of catalyst, about 4 drops would be required. While most catalysts also have a scale marking on their containers, desired quantities can also be poured out (usually by drops) by the change in levels on the scale. The markings are usually in ounces or cubic centimeters. Two hundred drops,

for example, equals about ⅙ ounce or 5 cubic centimeters of catalyst.

I find it most convenient to add the catalyst by counting drops when small amounts of resin are to be catalyzed. Most catalyst dispensers allow adding drops quickly, so 100 or more drops can be added in a short time period. If you intend to do a lot of fiberglassing, you will probably want to invest in a special catalyst dispenser that allows more rapid dispensing of desired amounts of catalyst. These are available from fiberglassing suppliers. When used, the catalyst is usually purchased in large containers and then poured into the special dispenser as needed. If the catalyst that comes with the resin you purchase does not come in a "drop" dispenser, or you purchase catalyst separately that doesn't, you can use a regular eye dropper for adding the catalyst by drops.

For the first practice fiberglassing exercises, only small amounts of resin will be catalyzed at a time. This can be applied quickly, so only a short pot life is required. Usually 2 to 3 percent catalyst by volume is needed to expedite the cure, which in turn will allow more practice fiberglassing in the same amount of time.

One ounce of resin is about 1/16 pint. One-half percent by volume of catalyst is about 2 drops; 1 percent is about 4 drops; 2 percent is about 8 drops; and 3 percent is about 12 drops. At 75 degrees Fahrenheit, the ½ percent would give a pot life of approximately 60 minutes. One percent would give approximately 30 minutes. Two percent would give about 15 minutes. Three percent would reduce the pot life or working time to about 7½ minutes, which would still allow ample time for applying this small amount of resin.

Using the same mixtures, higher working temperatures reduce the pot life and increase the curing rate. Lower working temperatures increase the pot life and decrease the curing rate.

General Principle

The general principle is the more catalyst, the faster the curing time; the less catalyst, the slower the curing time. Also, the higher the temperature, the faster the curing time; the lower the temperature, the slower the curing time. The same pot life and curing time can be maintained in a range of temperatures, usually from about 60 to 90 degrees Fahrenheit, with most regular polyester resins by varying the amount of catalyst used.

If at all possible, try to do your first practice fiberglassing exercises with a working temperature of about 75 degrees Fahrenheit, or at least at a temperature as close to this as possible. The idea is to keep the temperature variable constant at first so that you can better concentrate on other factors. After you have a little practice, you can learn to fiberglass in various temperatures.

Practice Exercise

The first exercise in catalyzing polyester resin demonstrates the effects of adding different amounts of catalyst to equal amounts of resin on gel and curing time. It gives an opportunity to study the characteristics of cured polyester resin without reinforcing material. You need either finishing or general-purpose polyester resin, catalyst for the resin, four mixing cups (the kind sold at drug stores with ounce markings on the sides are recommended), four small mixing sticks, and one longer mixing stick that can be used to stir the resin in the container in which it comes. You also need a piece of wax paper and paper to protect the area where you will be working, which can be on a bench, floor, or concrete surface outside. You should also have proper protective clothing and safety equipment (see Chapter 3). Follow the safety rules detailed in Chapter 4. This applies to all fiberglassing exercises and repair work covered in this book.

You also need a clock positioned so that it can be seen from the work area. You also need a pencil and notebook for recording the results of the practice exercises for future reference. Place a thermometer in the work area so that you can note the working temperature. You can start by writing the temperature down in your notebook.

When you have everything ready, begin by opening the container that the resin came in. Carefully stir it with a clean stirring stick. When removing the stick from the container, allow as much of

the resin to drain off the stick back into the container as possible. Use a clean rag to clean off the remaining resin from the stick. Put the stick aside for later use. Keep this one stick for use only for stirring uncatalyzed resin in the containers in which it is sold.

Pour 1 ounce of resin into each of the four cups. The markings on the sides of the cups can be used as a guide, or you can use a 1-ounce measuring scoop such as the kind used in the kitchen (but don't use it for kitchen duty again after you have used it for measuring out resin). Replace the lid on the resin container and set it aside.

To one of the cups containing 1 ounce of resin, add 2 drops of catalyst. Stir with one of the small sticks. Wait about 30 seconds. Observe the time on the clock and write this down in your notebook. Pour the catalyzed resin out on the piece of wax paper and spread it into a thin layer with the mixing stick. Use the small stirring stick to poke at the resin as the curing takes place. Because 2 drops of catalyst added to 1 ounce of resin is only about ½ percent by volume, it should take about an hour for the resin to gel. It should reach a cheeselike consistency and become hard. By using the stick as a probe, follow the process of the liquid changing to a solid. Write down how long it took from the time the catalyst was added until the gel state was reached, and then how much longer it took for the resin to become hard. The results depend on many factors, such as the particular resin and catalyst used, the working temperature, and the humidity. Almost certainly the gel time and curing time were much longer than really necessary.

With a second cup containing 1 ounce of resin, add 4 drops of catalyst or twice as much as was added to the first batch. This is about 1 percent by volume of catalyst. Again, mark down the starting time, noting the percentage of catalyst added. Again, observe the time to gel state and until the resin becomes hard. These times should be about half as long as previously. The resin should be poured from the cup to an area of the wax paper separate from the first (now hard) resin, which should be saved for later experimenting. Also, keep track of the percentage of catalyst that was added to each one.

Use the stirring stick to spread the resin out on the wax paper into a thin layer. This should be done so that the resin will still be in one piece when it hardens. Use the stick as a probe to determine when the gel state is reached and when the resin becomes hard.

With the third cup of 1 ounce of resin, repeat the same practice exercise, except this time add 8 drops of catalyst. This is approximately 2 percent by volume. Mark down the starting time. Stir. Wait about 30 seconds, then pour the resin out on the wax paper in an area away from the first two batches of resin, which should now be hard. Observe and record the time to gel and to when the resin becomes hard. For a typical polyester resin, the time to gel should be about 15 minutes, but your results might vary. It should become hard very soon—faster than was the case when less catalyst was used.

Data from the Exercise

With the last cup containing 1 ounce of resin, repeat the same steps, except this time add 12 drops of catalyst. This is about 3 percent by volume of catalyst. It should take only about 7½ minutes for the resin to reach the gel state. The resin should become hard very quickly. Record the times in your notebook. Your results should be similar to those shown in Table 7-1.

This will probably be enough fiberglassing for one practice session. Save the four pieces of cured resin. Mark them so that you can remember how much catalyst was added to each one. These pieces of cured resin will be used in the next practice session. Finish the first practice session by cleaning up the work area and putting everything away in an organized manner. Remember to store resin in a cool, dry place.

The data kept in your notebook will be useful as a reference for future fiberglassing. For example, if you later want to catalyze 4 ounces of the same resin, check your notebook. I made Table 7-1 when I carried out this practice exercise. From this table I can determine the amount of catalyst to be added for various working times. For instance, 16 drops of catalyst are used (4×4) for 30 minutes to gel or working time. Remember that the data is for a specific working temperature (my working temperature in this exercise was 75 degrees Fahrenheit). If you do future work at a different working temperature, the temperature will have to be taken into consideration and new test data gathered, as detailed previously in this chapter.

CURED POLYESTER RESIN WITHOUT REINFORCING MATERIAL

Polyester resin alone, without reinforcing material, is quite brittle when cured. Take the piece of resin that was formed by ½ percent by volume of catalyst and place it on a hard surface. Tap it with a hammer. It will probably shatter rather easily. Repeat with the other three pieces of cured resin. Could you observe any differences between the ease (or difficulty) of shattering them? In most cases you probably would not be able to detect any differences. The variations in amounts of resin used in this practice work probably would not have much affect on the strength of the cured resin. Things might be different if you went to greater extremes, such as less than ½ percent catalyst by volume or more than about 5 percent catalyst by volume. Too little catalyst would also give an unreasonably long working time, and too much catalyst would not give you enough time to apply the resin to anything.

The hammer test is a form of destructive testing. You found that the resin without reinforcing material shatters rather easily, but you also

Table 7-1. Data from the Exercise in Catalyzing Polyester Resin.

Amount of Resin (Oz.)	Percent of Volume of Catalyst	Drops of Catalyst	Min. to Gel	Min. until Hard
1	½	2	50	70
1	1	4	30	35
1	2	8	15	17
1	3	12	7-½	8

ruined the material in the process. This is an advantage of doing practice work before attempting actual fiberglassing repairs. On actual repair work, you will have little opportunity to do destructive testing. Destructive testing can tell you plenty about the strength and integrity of your fiberglass laminates that nondestructive testing can't, at least not so convincingly.

ADDING REINFORCING MATERIAL TO POLYESTER RESIN

Most of the same materials and supplies used for the first practice exercise will also be required for this one, except that you will only need one (clean) small mixing cup and one small stirring stick. For the following practice exercises, you should have an ample supply of mixing cups and stirring sticks on hand. For this exercise you also need some chopped strands of glass fibers. Because you will only need a small amount, and fiberglass mat will be required for future practice exercises, just purchase fiberglass mat and pull chopped strands of glass fibers from it. Save the rest of the mat for later. Store it inside a plastic bag.

Measure out an ounce of the general-purpose or finishing polyester resin into a mixing cup and then replace the lid on the resin container. Don't forget to stir the resin in the container before pouring or dipping out in a measuring cup the 1 ounce for use in this practice exercise. If this is not done, you will have thinner resin at the top of the container than at the bottom. The stirring assures an even consistency.

Before adding catalyst, add ⅓ ounce of chopped strands (either purchased chopped strands or those pulled from fiberglass mat) to the ounce of resin. With a small mixing stick, stir the resin and the chopped strands together. Make sure that all the chopped strands are saturated with resin.

Add 8 drops of catalyst to the resin. This should be about 2 percent by volume; you only count the resin, not the reinforcing material. You should have approximately 15 minutes of working time. Stir the catalyst into the resin. Then wait about 30 seconds. Pour the mixture out onto wax paper, using the stirring stick to get as much of it

from the cup to the wax paper as possible. Use the stick to spread the resin and reinforcing material into a layer about ⅛ inch thick.

Keep track of the time from when the catalyst was added until the gel state is reached and then the time until the mixture becomes hard. This should be approximately the same as when 8 drops of catalyst were added to 1 ounce of resin without reinforcing material.

After the mixture is hard and no longer tacky on the surface, pick it up from the wax paper. Note several things. First, you have more mass than when the resin alone was used previously. Second, the reinforced resin looks different than resin alone. You can probably see some of the strands of chopped glass fiber on the surface of the mixture. The side that was against the wax paper is probably quite smooth; the wax paper acted like a flat molding or forming surface.

Place the material on a hard surface and give it a tap with a hammer. Try to do this in the same manner as was done with the hard resin alone. Did the reinforced material shatter as easily? Unless something drastic happened, it should be much more difficult to shatter the reinforced resin. Even when it does break up, the reinforcing fibers tend to keep the material from shattering.

ADDING POLYESTER RESIN TO MAT

For this practice exercise you need, in addition to the general-purpose or finishing polyester resin, catalyst, and other basic items required for the earlier practice exercises, a ½-inch brush, a piece of cardboard, and a 6-inch square piece of 1½-ounce mat. Because you will need more mat later, purchase a larger piece and cut off a 3-inch square from one corner for use in this exercise. Put the remainder of the mat in a plastic bag, so it will stay clean and dry for later use.

Because 1 square foot of 1½-ounce mat weighs approximately 1½ ounces, the 6-inch square should weigh one-fourth as much, or about 0.38 ounce. Mat laminates are typically 25 to 35 percent fiberglass reinforcing material and 65 to 75 percent resin by weight. For the 25 percent mat ratio, about triple the weight of the mat in

59

resin is required to fully wet out the mat. For the 6-inch square of 1½-ounce mat, three times 0.38 ounce, or 1.14 ounce of resin, will be required.

For this practice exercise, pour about 1¼ ounces of resin into a mixing cup. You will need about 7½ minutes of working time. Look in your notebook and see how much catalyst you should add. For a working temperature of 75 degrees Fahrenheit, this should be about 12 drops for 1 ounce of resin or 15 drops for 1¼ ounces.

Before adding the catalyst, place the piece of mat on the cardboard (Fig. 7-1). Have wax paper ready to place the mat on once one side of the mat has been saturated with resin. Use the cardboard to turn the mat over, so the wet side of the mat will be face down on the wax paper. Have the brush ready.

Add the catalyst to the resin and stir. Wait about 30 seconds. Using the brush, apply resin to one side of the mat, which should be on the cardboard (Fig. 7-2). Use a dabbing action rather than a brushing motion so you do not lump up the mat.

This applies mainly to mat; you won't have the lumping problem with cloth and woven roving.

When the one side of the mat is fully saturated with resin, hold the cardboard, turn the mat over, and place it with the wet side down on the wax paper (Fig. 7-3). Using the brush, work any resin that remains on the cardboard onto the mat. Put the cardboard aside. Using the brush, dab the remaining resin on the other side of the mat, saturating the mat and getting the resin spread as evenly as possible (Fig. 7-4). When everything looks okay, clean the resin from the brush with acetone. This must be done before the resin hardens if the brush is to be used again.

Use a stirring stick to probe one corner of the mat to determine approximate gel time and curing time. After the laminate has cured (wait at least a few minutes after the laminate is hard), pick it up from the wax paper and examine it. If everything was done correctly, you should have a mat laminate made up of approximately 25 percent mat by weight and 75 percent resin by weight (Fig. 7-5).

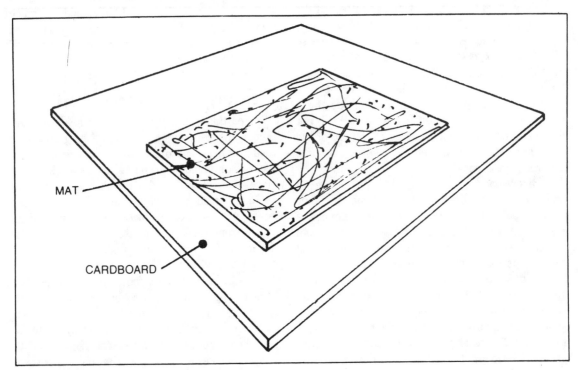

MAT

CARDBOARD

Fig. 7-1. Place the mat on a piece of cardboard.

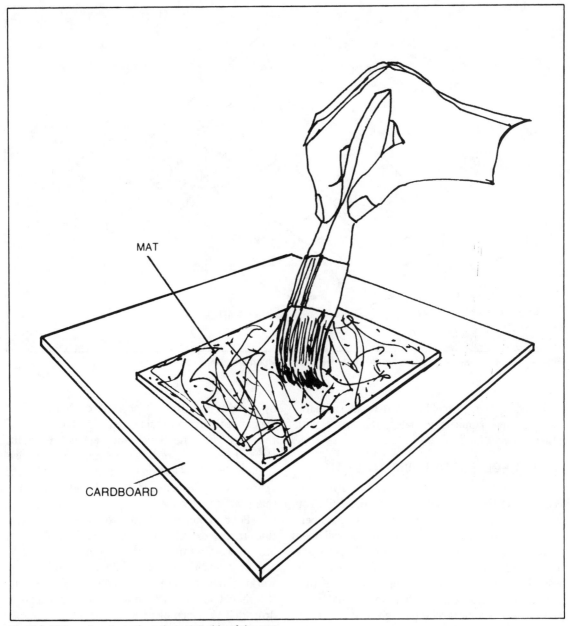

MAT

CARDBOARD

Fig. 7-2. Using a brush, apply resin to one side of the mat.

Slightly more resin was actually used, but some resin probably couldn't be recovered from the cardboard and brush. The finished laminate should weigh approximately 1½ ounces. The laminate should be about ⅟₂₀ inch thick. Also, the side of the mat that was placed over the wax paper should be quite smooth as the wax paper acted as a form or flat mold. The wax paper could have been placed on

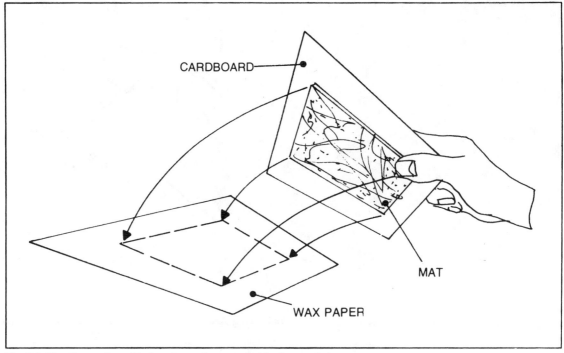

Fig. 7-3. Use the cardboard to turn the mat over wet side down on waxpaper.

a curved surface to mold a curved mat laminate. Save the mat laminate for later comparison with other laminates.

ADDING POLYESTER RESIN TO CLOTH

This is essentially the same as the mat exercise, except this time cloth is used (Fig. 7-6). You will need a 6-inch square piece of 10-ounce-per-square-yard fiberglass cloth. Because you will need more fiberglass cloth later, purchase a larger piece and cut off a 6-inch square from one corner. This will give you one selvaged edge and three that aren't. Be careful when cutting and handling the cloth so as not to unravel the unselvaged edges any more than necessary. Put the remainder of the fiberglass cloth in a plastic bag to keep it clean and dry for later use.

Because 1 square yard of 10-ounce cloth weighs approximately 10 ounces, the 6-inch square should weigh about 0.3 ounce, or $\frac{1}{36}$ as much. Cloth laminates are typically 45 to 50 percent fiberglass cloth by weight and 50 to 55 percent resin

by weight. For the 50 percent cloth ratio, the weight of the resin should be equal to the weight of the cloth. For the 6-inch square piece of cloth, about 0.3 ounce of resin will be required. Because some resin will be lost on the cardboard and brush, approximately ½ ounce of resin should be used for this practice exercise.

Pour ½ ounce of resin into a mixing cup. For this practice exercise, you will need about 7½ minutes of working time. Look in your notebook and see how much catalyst should be added. For a working temperature of 75 degrees Fahrenheit, this should be about 12 drops for 1 ounce resin or 6 drops for ½ ounce of resin.

Before adding the catalyst, place the piece of cloth on a piece of cardboard. Have wax paper ready as well as a brush for applying the resin.

Add the catalyst to the resin and stir. Wait about 30 seconds. Using the brush, apply resin to one side of the cloth, which should be on the cardboard. A brushing action can be used but take care not to unravel the unselvaged edges of the cloth.

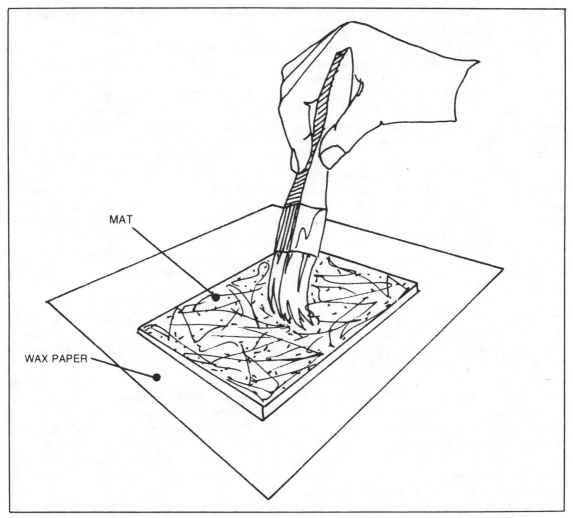

Fig. 7-4. Fully saturate the mat with resin.

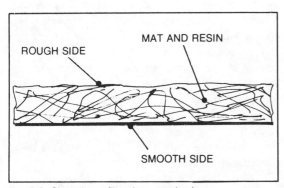

Fig. 7-5. Single-layer fiberglass mat laminate.

When one side of the cloth is fully saturated with resin, hold the cardboard, turn the cloth over, and place it with the wet side down on the wax paper. Using the brush, work any resin that remains on the cardboard onto the cloth. Put the cardboard aside. Brush the remaining resin onto the cloth. Saturate the cloth and get the resin spread as evenly as possible. When everything

Fig. 7-6. Single-layer fiberglass cloth laminate.

looks okay, clean the resin from the brush with acetone. This must be done before the resin hardens if the brush is to be used again.

Use a stirring stick to probe one corner of the cloth to determine approximate gel time and curing time. After the laminate has cured, pick it up from the wax paper and examine it. If everything was done correctly, you should have a cloth laminate made up of approximately 50 percent cloth by weight and 50 percent resin by weight. Slightly more resin was actually used, but some resin probably couldn't be recovered from the cardboard and brush. The finished laminate should weigh approximately 0.6 ounce and be about 1/64 inch thick, much thinner than the mat laminate. Compare it to the mat laminate to see if this is so. The mat laminate should be about three times as thick as the cloth laminate. The cloth laminate should weigh slightly more than one-third the weight of the mat laminate.

The side of the cloth that was placed over the wax paper should be relatively smooth because the wax paper acted as a form or flat mold. Depending on the viscosity of the resin used, how it was applied, and other factors, some of the cloth weave pattern probably shows through.

The other side of the cloth laminate is probably not as smooth and even as it was laid up without the use of a contact surface. With a single layer of cloth, a fairly even surface should result as long as too much resin isn't applied. Save the cloth laminate for later comparison with other laminates.

ADDING POLYESTER
RESIN TO WOVEN ROVING

This is essentially the same as the cloth exercise, except woven roving is used instead of cloth (Fig. 7-7). You need a 6-inch square piece of 24-ounce per square yard fiberglass woven roving. Because you will need more fiberglass woven roving later, purchase a larger piece and cut off a 6-inch square from one corner. Because woven

Fig. 7-7. Single-layer fiberglass woven roving laminate.

roving doesn't have selvaged edges, take care when handling and cutting the material to keep it from unraveling. Put the remainder of the woven roving in a plastic bag to keep it clean and dry for later use.

Because 1 square yard of 24-ounce woven roving weighs approximately 24 ounces, the 6-inch square should weigh about 0.68 ounces, or 1/36 as much. Woven roving laminates are typically 40 to 45 percent reinforcing material and 55 to 60 percent resin by weight. Thus, the weight of the resin will be slightly more than the weight of the woven roving. For the 6-inch square piece of woven roving, about 0.7 ounce of resin will be required. Because some of the resin will be lost on the cardboard and brush, use approximately 0.8 ounce of resin for this practice exercise. Pour 0.8 ounce of resin into a mixing cup.

This exercise needs about 7½ minutes of working time. A working temperature of 75 degrees requires about 12 drops of catalyst for 1 ounce of resin, or about 10 drops for the 0.8 ounce of resin.

Before adding the catalyst, place the woven roving on cardboard. Have wax paper and a clean brush for applying the resin ready.

Add the catalyst to the resin and stir with a mixing stick. Wait about 30 seconds. Using the brush, apply resin to one side of the woven roving, which should be on the cardboard. A brushing action can be used, but take care not to unravel the edges of the woven roving. This is a special problem with such a small piece of woven roving.

When one side of the woven roving is fully saturated with resin, pick up the cardboard, turn the woven roving over, and dump it wet side down on the wax paper. Using the brush, work any resin that remains on the cardboard onto the woven roving. Put the cardboard aside. Brush the remaining resin onto the woven roving. Saturate the woven roving and spread the resin as evenly over it as possible. When everything looks okay, clean the resin from the brush with acetone. This must be done before the resin hardens if the brush is to be used again.

Use a stirring stick to probe one corner of the woven roving to determine approximate gel time

64

and curing time. After the laminate has cured, pick it up from the wax paper and examine it. If everything was done correctly, you should have a woven roving laminate made up of approximately 45 percent woven roving by weight and 55 percent resin by weight. Slightly more resin was actually used, but some probably couldn't be recovered from the cardboard and brush.

The finished laminate should weigh about 1.4 ounces. The laminate should be about 1/25 inch thick, more than twice as thick as the cloth laminate and just slightly thinner than the mat laminate. Compare the woven roving laminate with those of cloth and mat to see if this is true. The woven roving laminate should weigh slightly less than the mat laminate and almost two and a half times as much as the cloth laminate.

The side of the woven roving that was placed over the wax paper should be relatively smooth, as the wax paper acted as a form or flat mold. The heavy weave pattern of the woven roving probably shows clearly. This is why mat is frequently combined with woven roving in laminates to fill in the weave pattern of the woven roving.

The other side of the woven roving laminate is probably not as smooth and even because it was laid up without the use of a contact surface. With a single layer of woven roving laid up over a flat surface, the noncontact surface should come out fairly even. Save the woven roving laminate for later comparison with other laminates.

SUMMARY OF PRACTICE LAMINATES USING MAT, CLOTH, AND WOVEN ROVING

These three practice exercises were intended to show some of the differences among mat, cloth, and woven roving when used with polyester resin in forming laminates. To make the results of the practice exercises more meaningful for future reference, they are multiplied out as though a *square foot* rather than a 6-inch square of each material was used. See Table 7-2.

Remember that the resulting laminates have different thicknesses. To produce laminates that are all approximately 3/32 inch thick, it takes two layers of 1 1/2 ounce mat, six layers of 10-ounce cloth, and 2.3 layers of 24-ounce woven roving.

While the total cost of the resin and reinforcing material for various laminates of the same thickness will vary depending on factors such as where and how you purchase your materials, it generally costs less for a mat laminate. The cost is about 25 percent more for a woven roving laminate, and the cloth laminate costs about twice as much as the woven roving. The cost of the resin is nearly the same for all three laminates. The resin for mat costs the most, followed by the resin for cloth, and then the resin for woven roving. It's mainly the cost of the reinforcing materials that makes the difference. Mat is the least expensive. Cloth is about four times as expensive as mat. Woven roving is about half the cost of the cloth, or about 80 percent higher than mat. Remember that these figures are for the same total thickness of laminate.

Why not just use mat and save money? Whenever mat will serve the purpose, this is the economical way to go. Cloth and woven roving have important strength advantages over mat, so there are many fiberglassing repair jobs where these reinforcing materials should be used.

So far, laminates of only a single layer of one

Table 7-2. Results of Practice Exercises, Assuming a Square Foot of Each Material Was Used.

	1½ Oz. Mat	10 Oz. Cloth	24 Oz. Woven Roving
Weight of Reinforcing Material (Oz.)	1.5	1.2	2.7
Weight of Resin (Oz.)	4.6	1.2	2.8
Weight of Laminate (Oz.)	6.1	2.4	5.5
Thickness of Laminate (inches)	1/20	1/64	1/25

kind of reinforcing material have been used in the practice exercises. Now you will experiment with laminates with two or more layers of the same reinforcing material.

LAMINATING TWO LAYERS
OF MAT WITH POLYESTER RESIN

For this practice exercise, essentially the same materials will be required as were used for the single-layer mat laminate (Fig. 7-8). This time, though, you need two 6-inch square pieces of 1½-ounce mat instead of one. You could add another layer to the original mat laminate, except you will need the single-layer one for comparison with the two-layer one. Also, if finishing resin was used, the wax would have to be removed from the surface chemically or by sanding to get a good bond. For this practice exercise you will need either general-purpose polyester resin or both laminating and finishing resin. To better understand the differences between laminating (air-inhibited) and finishing (nonair-inhibited) resins, I highly suggest that these be used for at least some of the practice work.

Begin by making a single-layer mat laminate as was done previously, except this time use either general-purpose or laminating resin. Allow the laminate to cure. If laminating resin is used, the resin should have a tacky surface if the cure takes place in the presence of air. The resin should not contain wax to shield the air from the surface, so the surface can fully cure. General-purpose resin generally contains some wax to allow a reasonable surface cure but not so much as to cause laminating problems. Some fiberglassing experts disagree with this point, believing that general-purpose resins don't do either job very well.

Place the single-layer mat laminate on the wax paper. It would be possible to lay up two layers of mat at once, without allowing the resin from the first layer to cure before adding the second. This

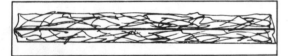

Fig. 7-8. Two-layer fiberglass mat laminate.

can add complications, so wait until later to try it.

Because the next layer of mat will be the last layer to be added to this laminate, use laminating, finishing, or general-purpose resin. The wax on the cured surface causes most of the bonding problems, not the wax in the new resin that is being applied. If additional layers are to be added later to the laminate, use laminating resin. If laminating resin is used for this practice exercise, apply a thin layer of finishing resin to both sides of the laminate after the laminating resin has cured, so the surface will cure tack-free.

This might seem confusing at first. After you have worked with the different types of resin for a while, though, everything will become clear.

Place the mat to be used for the second layer on cardboard. Wet out the upper surface with catalyzed resin. Brush on a thin layer of resin on the cured laminate that is on the wax paper. Turn the cardboard over and place the wet resin sides of the two layers together. With the brush, use a dabbing motion to work the two layers together. Wet out the second layer of mat until it is fully saturated with resin. When everything looks okay, clean the brush with acetone before the resin has a chance to harden.

If laminating resin was used, wait until the surface has cured to a tacky state. The resin below the surface will be cured by this time. The surface layer keeps air from the resin underneath and allows it to cure. Catalyze a small quantity of finishing resin and brush on a thin layer to both sides of the laminate. The side that was previously on the wax paper can be placed back on the wax paper on the same side as soon as that side has been coated with finishing resin. Often the wax paper itself will inhibit the air from that side of the laminate and allow the surface to cure tack-free, even when laminating resin is used.

If finishing resin was used throughout, the wax could be removed from the surface of cured resin with acetone. Put the acetone on a clean white cloth and wash the surface with acetone. The wax can also be removed by sanding the surface, but this is more difficult.

Examine the laminate with two layers of mat. It should be approximately twice as thick as the previous mat laminate that had only one layer of mat. With a sharp object, try to separate the laminate at one edge between the two layers of mat. If everything was done correctly and quality materials were used, the laminate should show little or no tendency to delaminate here.

As more and more layers are added to a laminate, the back side of the laminate tends to become more and more uneven, especially when mat is used. Some bunching up and lumping of the mat is almost inevitable.

LAMINATING THREE LAYERS OF MAT WITH POLYESTER RESIN

Repeat the same practice exercise, but make a laminate with three layers of 1½-ounce mat. Do this in three stages. Allow the first layer of the laminate to cure before adding the second layer. Allow this layer to cure before adding the third layer. Allow the laminate to cure.

The three-layer laminate should be approximately ⅛ inch thick. Compare this with the two-layer laminate, which should be about 3/32 inch thick. The single-layer laminate should be about 1/20 inch thick.

LAMINATING TWO LAYERS OF CLOTH WITH POLYESTER RESIN

This practice exercise requires essentially the same materials as for the laminate with two layers of mat. The exception is that 10-ounce fiberglass cloth will be used instead of mat.

Make a single-layer cloth laminate, except use either general-purpose or laminating resin. Allow the laminate to cure. If laminating resin is used, the resin should have a tacky surface if the cure takes place in the presence of air. The resin should not contain wax to shield the air from the surface, so the surface can fully cure. Leave the single-layer laminate on the wax paper.

If this is the last layer to be added to this laminate, use laminating, finishing, or general-purpose resin. If additional layers are to be added to the laminate later, use laminating resin. If laminating resin is used for this practice exercise, apply a thin layer of finishing resin to both sides of the laminate after the laminating resin has cured, so the surface will cure tack-free.

Place the cloth for the second layer on cardboard. Wet out the upper surface with catalyzed resin. Brush on a thin bonding layer of resin to the cured laminate that is on the wax paper. Turn the cardboard over and place the wet resin sides of the two layers together. Line them up so that the edges are even. Use the brush to smooth out the upper layer of cloth. Wet out the upper layer of cloth until it is fully saturated with resin. Work out any air bubbles that appear in the cloth by dabbing them down with the brush and working them off to one side of the laminate. When everything looks okay, clean the brush with acetone before the resin has a chance to harden.

If laminating resin was used, wait until the surface has cured to a tacky state. The resin below the surface will be cured by this time. Catalyze a small quantity of finishing resin and brush on a thin layer to both sides of the laminate. The side that was previously on the wax paper can be placed back on the wax paper on the same side as soon as that side has been coated with finishing resin. The wax paper can inhibit the air from that side of the laminate and allow the surface to cure tack-free, even when laminating resin is used.

If you should later decide to add more layers to a laminate with finishing resin on the surface, remove the wax from the surface with acetone. Apply acetone to a clean white cloth and wash the surface. The wax can also be removed by sanding the surface.

Examine the laminate with the two layers of cloth. It should be approximately twice as thick as the one-layer cloth laminate. With a sharp object, try to separate the laminate at one edge between the two layers of cloth. If everything was done correctly and quality materials were used, the laminate should show little or no tendency to delaminate here.

The two-layer cloth laminate is thinner than the single-layer mat laminate. While cloth has many characteristics that make it useful for certain types of fiberglass laminating work, it isn't good for building up laminate thickness. The use of cloth is also an expensive method for building up thickness as cloth is much more expensive than mat or woven roving for achieving the same thickness in a laminate.

The back side of the cloth laminate should be fairly smooth, though not as smooth and even as the side that was against the wax paper. Cloth tends to remain fairly even unlike mat, which tends to bunch up in some areas and form an uneven surface.

LAMINATING THREE LAYERS OF CLOTH WITH POLYESTER RESIN

Repeat the same practice exercise, except this time add an additional third layer of cloth to the laminate. Do this in three stages. Allow the first layer of the laminate to cure before adding the second layer. Allow this layer to cure before adding the third layer.

Allow the laminate to cure. If laminating resin was used, add a thin layer of finishing resin to the surfaces of the laminate, so it will cure with a tack-free surface.

The three-layer cloth laminate should be about 1/20 inch thick, about the same as a single-layer mat laminate made with 1½-ounce mat. The three-layer cloth laminate is stronger (assuming that the laminate was laid up properly and quality materials were used). The materials for the cloth laminate were more expensive. It took approximately three times as long to lay up because three layers of reinforcing material were required instead of one. The cloth laminate contains more fiberglass reinforcing material and less resin.

LAMINATING TWO LAYERS OF WOVEN ROVING WITH POLYESTER RESIN

This practice exercise requires the same materials as for the laminate with two layers of cloth. Instead of 10-ounce cloth, use 24-ounce fiberglass woven roving.

Make a single-layer woven roving laminate using either general-purpose or laminating resin. Allow the laminate to cure. If laminating resin is used, the resin should have a tacky surface if the cure takes place in the presence of air. The resin should not contain wax to shield the air from the surface, so the surface can fully cure. Because this is the last layer of woven roving to be added to this laminate, use laminating, finishing, or general-purpose resin.

Place the woven roving for the second layer on cardboard. Wet out the upper surface with catalyzed resin. Brush on a bonding layer of resin to the cured laminate that is on the wax paper. Turn the cardboard over and place the wet resin sides of the two layers together. Line them up so that the edges are even. Use the brush to smooth out the upper layer of woven roving, taking care not to unravel the weave of the woven roving. Wet out the upper layer of woven roving until it is fully saturated with resin. When everything looks okay, clean the brush with acetone before the resin has a chance to harden.

If laminating resin was used, wait until the surface has cured to a tacky state. The resin below the surface has cured by this time. The surface layer keeps air from the resin underneath and allows it to cure. Catalyze a small quantity of finishing resin and brush on a thin layer to both sides of the laminate. Clean the brush with acetone. Allow the surface of the laminate to cure tack-free.

Examine the laminate. It should be approximately twice as thick as the previous woven roving laminate that had only one layer of woven roving. With a sharp object, try to separate the laminate at one edge between the two layers of woven roving. If everything was done correctly and quality materials were used, the laminate should show little or no tendency to delaminate. Woven roving tends to have a resin-rich area between layers when laminated without a mat layer to fill in the coarse weave pattern of the woven roving. The two-layer woven roving laminate should be only slightly thinner than the two-layer mat laminate.

LAMINATING THREE LAYERS OF WOVEN ROVING WITH POLYESTER RESIN

Repeat the same practice exercise, except this time add an additional third layer of woven roving to the laminate. Do this in three stages. Allow the first layer of the laminate to cure before adding the second layer. Allow this layer to cure before adding the third layer.

Allow the laminate to cure. If laminating resin was used, add a thin layer of finishing resin to the surfaces of the laminate, so it will cure with a tack-free surface.

The three-layer woven roving laminate should be about ⅛ inch thick, slightly thinner than the three-layer mat laminate. The three-layer woven roving laminate is much stronger (assuming that the laminate was laid up properly and quality materials were used). Woven roving also builds up thickness quickly and is only slightly more expensive on a laminate thickness basis than mat. Woven roving is a fairly difficult material to work with, however. For laminating work, a layer of mat is generally sandwiched between layers of woven roving.

OTHER LAMINATING METHODS

In the previous mat, cloth, and woven roving exercises, the reinforcing material was placed on a piece of cardboard and wetted out on one side first, before being placed on the wax paper or added to a laminate. This method works well for small pieces of reinforcing material, but other methods also work. For example, a layer of catalyzed resin could be applied to the wax paper and then dry reinforcing material placed on it. When additional layers are added to a laminate, the same technique can be used. Brush on a layer of resin to the surface of the cured laminate where the next layer of laminate is to be added. Place the dry reinforcing material on the wet resin. Saturate the reinforcing material with catalyzed resin. This method will usually suffice to adequately wet out the reinforcing material.

I suggest that you try this method of laminating by first using mat, then cloth. Compare the resulting laminates with the ones laid up by using the cardboard wet-out method to see which method works best for you.

Still another method is to apply two or more layers of reinforcing material in one operation, without allowing the first to cure before adding additional layers. This can result in a faster lay-up, but it can result in problems, especially when laying up large pieces of reinforcing material. Also, there is a limit to the number of layers that can be added

in one operation, usually two or three, while still maintaining a quality lay-up.

As a practice exercise, try laying up two layers of mat in one operation. Begin by placing the first layer of mat on a wet layer of resin on the wax paper. Saturate the mat with resin. Use a dabbing action of the brush to keep the mat from lumping up. Place the second layer of mat on the wet resin of the first layer. Finish wetting out the second layer of mat. When this laminate has cured, compare it with a laminate with two layers of mat that was laid up with the first layer allowed to cure before the second was added.

Repeat this same practice exercise using two layers of cloth. Place the first layer of cloth on a wet layer of resin on the wax paper. Saturate the cloth with resin. Place the second layer of cloth on the wet resin of the first layer. Finish wetting out the second layer of cloth. When this laminate has cured, compare it with a laminate with two layers of cloth that were laid up with the first layer allowed to cure before the second was added.

From these practice exercises, it should be apparent that there is no method that is best for laying up a laminate. Experiment to find out what method works best for you. Some methods are faster than others, but it's important that you don't lose control. Also, it is fairly easy to lay up two layers of reinforcing material in one operation when small pieces of material are used, but this can be much more difficult to do without losing control when large pieces of reinforcing material are used.

For these practice exercises, the lay-up work was done with a brush. You might now want to use squeegees and laminating rollers. These will be especially useful when working with fairly large pieces of reinforcing material.

LAMINATES WITH MORE THAN ONE KIND OF REINFORCING MATERIAL

For many types of fiberglassing repair work, it is desirable to use more than one type of reinforcing material in the same laminate. The importance of using a mat layer between layers of woven roving has always been explained. Mat can also be effectively combined with cloth. In some laminates mat, cloth, and woven roving will be used. Composite laminates are used to take advantage of the characteristics of the different reinforcing materials. For example, mat gives thickness and fills in small spaces such as the weave pattern of cloth and woven roving. Mat also gives a smooth surface against a mold. Cloth adds strength to a laminate but doesn't give much thickness, and it is also expensive.

After doing the practice exercises, you have probably found a laminating method that works best for you. This can be used for the following practice exercises using more than one kind of reinforcing material in the same laminate.

Mat and Cloth

Using 6-inch squares of reinforcing material, make a laminate of one layer of 1½-ounce mat and one layer of 10-ounce cloth (Fig. 7-9). Allow the laminate to cure. With a sharp object, try to separate the two layers of the laminate at one edge. If everything was done properly and quality materials were used, there is a strong bond between the layers of reinforcing material. The thickness of the laminate should be equal to approximately the combined thickness of a one-layer mat laminate and a one-layer cloth laminate with the same weights of reinforcing materials. The laminate should appear different on the two surfaces. Short strands of reinforcing material should show on the mat side. The weave pattern of the cloth should show on the cloth side.

Make a cloth-mat-cloth laminate. The mat is sandwiched between two layers of cloth. This type of laminate tends to smooth out the surfaces of the mat on the side of the laminate away from the mold

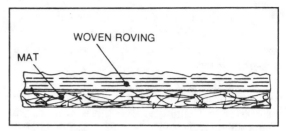

Fig. 7-10. Laminate with one layer of mat and one layer of woven roving.

or form, but it leaves the weave of the cloth showing on the surface.

Make a mat-cloth-mat-laminate. The cloth will be sandwiched between two layers of mat. This adds considerable strength to a two-layer mat laminate without adding much thickness. The cloth is expensive, however, so this must be taken into consideration. Many fiberglassing repair jobs call for even thicker laminates of mat and cloth in various combinations.

Mat and Woven Roving

Using 6-inch squares of reinforcing material, make a laminate of one layer of 1½-ounce mat and one layer of 24-ounce woven roving (Fig. 7-10). While various laminating techniques can be used, it is best to place the woven roving on wet mat or to place the mat on wet woven roving so that the mat can fill in the coarse weave of the woven roving. If one layer is allowed to cure first, added reinforcing material will not penetrate the surface. The reinforcing materials must be thoroughly saturated with resin. Allow the laminate to cure.

The woven roving adds considerable thickness to the laminate, almost as much as the mat, and greatly increases the strength as compared to a laminate of mat only. For the same thickness, the woven roving layer of the laminate (woven roving and resin) is only slightly more costly than the materials for the mat layer.

With a sharp pointed metal object, try to separate the two layers of the laminate at one edge. If everything was done properly and quality materials were used, there is a strong bond between the layers of reinforcing material. The thickness of the

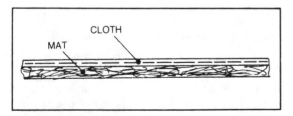

Fig. 7-9. Laminate with one layer of mat and one layer of cloth.

laminate equals approximately the combined thickness of a one-layer mat laminate and a one-layer woven roving laminate with the same weights of reinforcing materials. The laminate again appears different on the two surfaces. Short strands of reinforcing material show on the mat side. The heavy weave pattern of the woven roving shows on the woven roving side.

Make a woven roving-mat-woven roving laminate. The mat is sandwiched between two layers of woven roving. This lay-up should fill in the coarse weave pattern of the woven roving layers inside the laminate but leave the coarse weave pattern showing on the outside surfaces.

Make a mat-woven roving-mat laminate. The woven roving will be sandwiched between two layers of mat. This adds considerable strength to a two-layer mat laminate and increases the thickness by almost one third.

When woven roving is used in laminates, it is frequently alternated with layers of mat. If this is not done, pockets of resin without reinforcing material tend to form in the coarse weave pattern of the woven roving. This results in resin-rich areas that are brittle and tend to crack under stress.

Mat, Cloth, and Woven Roving

Using 6-inch squares of reinforcing material, make a laminate of one layer of 1½-ounce mat, one layer of 10-ounce cloth, another layer of 1½-ounce mat, and a layer of 24-ounce woven roving (Fig. 7-11). Use wax paper as a flat form or mold for the first layer of mat. While various laminating methods can be used, I suggest that the first layer of mat be laid up and allowed to cure. Then add the cloth layer. Allow this to cure. Next, add both the mat and woven roving layers all in one operation, so the mat will fill in the coarse weave pattern of the woven roving.

Allow the laminate to cure. The laminate should be smooth on the side that was against the wax paper. The other side should show the coarse weave pattern of the woven roving. The laminate gives a good combination of thickness and strength, assuming that the lay-up was done properly and quality materials were used. Laminates

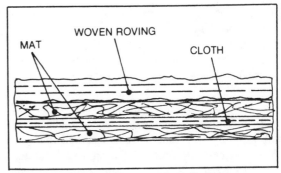

Fig. 7-11. Mat, cloth, and woven roving laminate.

with combinations of reinforcing materials are frequently used for fiberglassing repairs.

SUMMARY OF TECHNIQUES OF LAMINATING WITH POLYESTER RESIN

The practice exercises were designed for learning the basic techniques of forming laminates with various reinforcing materials and polyester resin. If you have conscientiously performed the practice exercises, you learned the basics of catalyzing polyester resins and applying the resin to mat, cloth, and woven roving to form laminates.

You now routinely follow the safety rules and precautions given in Chapter 4 and on the labels of the materials used. You know the importance of careful working habits and clean working conditions. Because you are doing a chemical form of construction, you keep the work areas clean to avoid contaminating the chemicals being used.

You now realize that the polyester resin must not only be applied to the reinforcing material but in the proper amount depending on the reinforcing material used. Mat laminates generally contain about 25 to 30 percent mat reinforcing material by weight, with the remaining 70 to 75 percent resin by weight. Cloth laminates contain about 45 to 50 percent cloth reinforcing material by weight, with the remaining 50 to 55 percent resin by weight. Woven roving laminates contain about 40 to 45 percent woven roving reinforcing material by weight, with the remaining 55 to 60 percent resin by weight.

Keep notes relating to fiberglassing on the amount of catalyst added, working time, the resin required for various types of reinforcing materials, and so on. These notes are valuable for future reference.

You have begun to develop your own working methods. There is no *one* right way to do fiberglassing.

Additional practice exercises will be presented in this chapter to further develop your fiberglassing skills and techniques. It is assumed that you have mastered the previous techniques. In the following exercises it will be assumed that you will catalyze the polyester resin before adding it to reinforcing material, that you will add an amount of catalyst that will give you sufficient working time for the particular job, and that you will know the proper amount of resin to use for each type of reinforcing material.

ADDING A MAT AND
POLYESTER RESIN LAMINATE TO WOOD

In addition to the basic materials required for the previous practice exercises, you need a 6-inch square piece of plywood (Fig. 7-12). This can be cut from ¼, ⅜, or ½-inch plywood. Exterior plywood of any grade can be used. Avoid interior grade plywood for fiberglassing work. You also need sandpaper in an assortment of grits from coarse to fine and a sanding block (see Chapter 3).

In this and following exercises, a layer of fiberglass (resin and reinforcing material) will be laid up on and bonded to plywood. This can be thought of in many ways, such as wood covered on one side with a layer of fiberglass or as a composite laminate of plywood and fiberglass. The fiberglass laminate will be chemically laid up and bonded to the plywood. This is different from, say, gluing or other-

wise bonding a cured sheet of fiberglass laminate to plywood, which is also a possibility similar to adding plastic laminates to plywood for tabletops.

It is sometimes difficult to get a good bond between the laminate and plywood using polyester resin. The more expensive epoxy resin gives a better bond. This should become apparent later when practice exercises are done using epoxy resin.

For this exercise, take one of the 6-inch squares of plywood and, using coarse sandpaper, rough sand one side of the plywood. This will help give the resin a better bond. After the surface of the wood has been rough sanded, wipe the area down with an acetone-soaked rag. While opinions vary on how the wood should be sanded for fiberglassing, I have found it best to use cross-grain sanding and to scratch up the surface of the wood considerably while at the same time keeping the surface of the wood fairly even.

Using scissors or snips, cut a 6-inch square of 1½-ounce mat. Remember to wear polyethylene gloves to protect the skin and to help keep the mat clean.

Place the plywood on heavy paper arranged to protect the work area. Pour enough resin for wetting out the mat into a mixing cup. This will be approximately the same amount of resin that was required for a one-layer mat laminate, about 1¼ ounces. Finishing, general-purpose, or laminating resin can be used. If laminating resin is used, a small amount of finishing resin must be added to the surface of the laminate to help it cure tack-free.

Add catalyst to the resin in the mixing cup. Approximately 7½ minutes of working time will be about right. Stir catalyst into the resin. Wait about 30 seconds.

Using a brush, apply a layer of resin to the plywood surface where the laminate is to be added. Place the dry mat in position on the plywood in the wet resin. An alternate method is to place the mat on cardboard and wet out one side of the mat. This is then turned over onto the wet resin on the plywood.

Apply resin to the mat with the brush. Use a dabbing, rather than brushing, action to avoid bunching the mat up. Once the mat is thoroughly

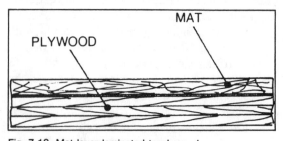

Fig. 7-12. Mat layer laminated to plywood.

saturated with resin, do not apply additional resin. For best results, no more than .75 percent of the laminate by weight should be resin.

Allow the resin to cure tack-free. If laminating resin was used, a thin layer of finishing resin should be added to the surface of the cured laminating resin, so the surface will cure tack-free.

Examine the laminate. Using a sharp-pointed metal object, try to separate the fiberglass laminate from the plywood at one edge. The effectiveness of the bond depends on many factors. If proper techniques and quality materials were used, a reasonably strong bond should have resulted.

Save the laminate for comparison with other laminates. These laminates will also be useful later for practice sanding—an important skill for fiberglassing repair work.

ADDING A CLOTH AND POLYESTER RESIN LAMINATE TO WOOD

This practice exercise is essentially the same as the previous exercise, except that cloth will be used instead of mat (Fig. 7-13). Using scissors or snips, cut a 6-inch square of 10-ounce fiberglass cloth. Remember to wear polyethylene gloves to protect your skin and to help keep the cloth clean.

Place a 6-inch square piece of plywood on heavy paper arranged to protect the work area. Pour enough resin for wetting out the cloth into a mixing cup. Add catalyst.

Using a brush, apply a layer of resin to the plywood surface where the laminate is to be added. Place the dry cloth in position on the plywood in the wet resin. An alternate method is to place the cloth on cardboard and wet out one side of the cloth. This is then turned over onto the wet resin on the plywood.

Apply resin to the cloth with the brush. Once

the cloth is thoroughly saturated with resin, do not apply additional resin. For best results, no more than 55 percent of the laminate by weight should be resin. Smooth out the cloth and remove any air bubbles that form. This can be done with the brush, or a squeegee can be used.

Allow the resin to cure tack-free. If laminating resin was used, a thin layer of finishing resin should be added to the surface of the cured laminating resin to help the surface cure tack-free.

Examine the laminate. Using a sharp-pointed metal object, try to separate the fiberglass laminate from the plywood at one edge. The effectiveness of the bond depends on many factors. If proper techniques and quality materials were used, a reasonably strong bond should have been made. Save the laminate for comparison with other laminates and for practice sanding.

ADDING A WOVEN ROVING AND POLYESTER RESIN LAMINATE TO WOOD

This practice exercise is essentially the same as the previous one, except that woven roving will be used instead of cloth (Fig. 7-14). Using scissors or snips, cut a 6-inch square of 24-ounce fiberglass woven roving.

Place a 6-inch square piece of plywood on heavy paper arranged to protect the work area. Pour enough resin for wetting out the woven roving into a mixing cup. Add catalyst.

Using a brush, apply a layer of resin to the plywood surface where the laminate is to be added. Place the dry woven roving in position on the plywood in the wet resin. An alternate method is to place the woven roving on cardboard and wet out one side of the woven roving. Then turn it over

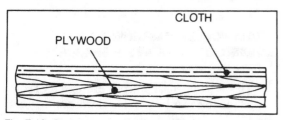

Fig. 7-13. Cloth layer laminated to plywood.

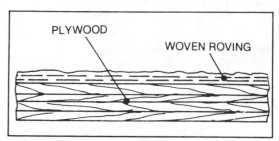

Fig. 7-14. Woven roving layer laminated to plywood.

onto the wet resin on the plywood. Remember, however, that the latter method is generally only suitable for use with fairly small pieces of reinforcing material.

Apply resin to the woven roving with the brush. Once the woven roving is thoroughly saturated with resin, do not apply additional resin. For best results, no more than 60 percent of the laminate by weight should be resin. Smooth out the woven roving and remove any air bubbles that form. This can be done with the brush or a squeegee. When the correct amount of resin is used, the heavy weave pattern will still clearly show. The tendency might be to fill in the low areas with resin. This will result in areas of resin without reinforcing material that will be brittle and subject to cracking.

Allow the resin to cure tack-free. If laminating resin was used, add a thin layer of finishing resin to the surface of the cured laminating resin.

Examine the laminate. Using a sharp-pointed metal object, try to separate the fiberglass laminate from the plywood at one edge. If proper techniques and quality materials were used, a reasonably strong bond will result.

COMPARISON OF MAT, CLOTH, AND WOVEN ROVING LAMINATES BONDED TO WOOD

Compare the mat, cloth, and woven roving laminates bonded to wood. The mat should have given the thickest laminate, with the woven roving just slightly thinner. The cloth provided a very thin laminate.

Assuming that the pieces of plywood were equally rough sanded for each laminate, the mat would typically give the best bond to the wood. The cloth would also provide a good bond, though not quite as good as mat. Woven roving is more difficult to bond to wood, although it can be done adequately for some purposes.

The main problem is that polyester resin does not bond very well to wood or, for that matter, to most other materials. The polyester resin does not saturate very far into the surface of the wood. Test this out on one of the practice laminates by using a

sharp-pointed metal object to chip away some resin from the wood.

There are some ways to overcome this problem. The reinforcing material can be mechanically fastened to the wood with staples or by other means. The more expensive epoxy resin, which gives a superior bond, can be used instead of polyester resin.

Another problem is that fiberglass and wood have different physical properties, which in some cases can cause the fiberglass to delaminate from the wood. For this reason, nonfiberglass reinforcing materials such as polyester, which has greater elasticity, are sometimes used instead of fiberglass reinforcing material for covering wood.

To staple reinforcing material to wood, both the reinforcing material and the wood must be dry. Resin will quickly clog up most staple guns, and most staples cannot be driven through cured resin. When the reinforcing material is stapled dry to the wood, resin applied to the outside of the reinforcing material will not provide a good bond with the wood for most reinforcing material. A large open weave cloth seems to work best. Polypropylene and some other nonfiberglass reinforcing materials can be stapled on dry, adequately wetted out, and bonded by applying polyester resin from the outside of the material.

The three laminates each gave different exterior surfaces. The mat has a tendency to bunch up in some areas, giving a somewhat uneven surface. The cloth usually gives a smooth surface, but the weave pattern of the material tends to show through. This can be minimized by using lightweight cloth with a close weave pattern. The coarse weave pattern really shows through on woven roving and makes it very difficult to sand smooth, unless additional reinforcing material and resin are applied to fill in the weave pattern.

MECHANICALLY FASTENING AND BONDING OPEN WEAVE FIBERGLASS CLOTH TO WOOD

This is an optional practice exercise for those who intend to cover large areas of wood with fiberglass, such as covering a plywood boat with a fiberglass skin. In addition to the materials for the previous practice exercises, you will need fiberglass cloth with a large open weave pattern (7.5 ounces per square yard is about right), a heavy-duty staple gun, and staples. While regular steel staples can be used, rustproof *monel staples* are recommended, especially for boats and other marine applications.

Rough sand one surface of a 6-inch square piece of plywood for bonding. Wash with acetone.

Using scissors or snips, cut out a 6-inch square of cloth with a large open weave pattern. Place the cloth dry on the wood and staple it to the wood. Space the staples a couple of inches apart, either in a pattern or at random. Make sure that the cloth is stretched tight and smooth.

Place the plywood with the fiberglass cloth stapled to it on heavy paper arranged to protect the work area. Pour enough resin for wetting out the cloth into a mixing cup. Add catalyst.

Using a brush, saturate the cloth with resin. Once the cloth is thoroughly saturated with resin, do not apply additional resin. A common mistake is to apply too much resin.

When everything looks okay, clean the brush with acetone before the resin has a chance to harden. Allow the laminate to cure.

The laminate should be both bonded and mechanically fastened to the plywood. Set the staples below the surface of the laminate. Additional layers of reinforcing material can be added to this by using regular laminating methods.

MECHANICALLY FASTENING AND BONDING POLYESTER CLOTH TO WOOD

This is another optional practice exercise for those who intend to cover large areas of wood with fiberglass (I'm using the term fiberglass even though no glass fiber reinforcing materials are used), such as a plywood boat. For this practice exercise, you will need a 6-inch square of 4.2-ounce-per-square-yard polyester reinforcing material. Purchase a larger piece of the material and cut off a 6-inch square. Place the remainder of the material in a plastic bag for storage.

Rough sand one surface of a 6-inch square piece of plywood. Wash the surface with acetone. Allow the surface of the wood to dry.

Place the dry polyester cloth on the wood and staple it in place. Space the staples a couple of inches apart, either in a pattern or at random. Make sure that the cloth is stretched tight and smooth.

Place the plywood with the polyester cloth stapled to it on heavy paper arranged to protect the work area. The 4.2-ounce polyester cloth requires about the same amount of resin as 10-ounce fiberglass cloth. Pour the resin into a mixing cup. Add catalyst. Using a brush, saturate the cloth with resin. This will give a soak-through bond to the wood. Once the cloth is thoroughly saturated with resin, do not apply additional resin. A common mistake is to use too much resin. It is far better to apply too little resin, as more can always be added later. It is very difficult to remove excess resin.

When everything looks okay, clean the brush with acetone before the resin has a chance to harden. Allow the laminate to cure.

Examine the laminate. Even without the staples, polyester cloth bonds well with polyester resin to wood. The staples supplement the bond and allow the polyester cloth to be held in position dry. This is a big advantage when covering large areas of wood.

Using a sharp-pointed metal object, test the bond to the wood at one edge. Part of the wood will usually break away before the bond will separate.

Additional layers of polyester cloth can be added to the laminate. This is done without the use of staples, as it is almost impossible to drive them through the first cured layer of the laminate.

While opinions vary, I prefer to use polyester cloth rather than fiberglass cloth for covering plywood. The material is only slightly more expensive than fiberglass and, in my opinion, well worth the small additional cost. Polyester cloth is an extremely easy material to wet out with polyester resin. The resulting laminate has greater adhesion, toughness, and abrasion resistance than a similar laminate using fiberglass cloth. Unlike fiberglass reinforcing material, the polyester cloth is nonirritating to the skin, an important advantage especially for those who have dermatological sensitivity to working with fiberglass reinforcing materials.

MAT AND POLYESTER
RESIN LAMINATE WITH A CORE

Some fiberglassing repair work calls for laminates with core materials, such as end-grain balsa or rigid plastic foams. For this practice exercise, you will need a 6-inch square of ¼-inch thick polyurethane rigid foam or closed-cell polyvinyl chloride foam. Polystyrene foam cannot be used with polyster resin.

A two-layer mat laminate will be added to each side of the core material to form a sandwich core laminate (Fig. 7-15). Cut out four 6-inch squares of 1½-ounce mat.

Place the core material on heavy paper arranged to protect the work area. Pour enough resin for wetting out one layer of mat into a mixing cup. Add catalyst.

Using a brush, apply an even coat of resin to one side of the core material. Place a layer of dry mat on the resin and press it into the wet resin. Apply additional resin to the mat by using a dabbing action with the brush so you do not bunch up the mat. Thoroughly saturate the layer of mat with resin. When everything looks okay, clean the brush with acetone before the resin has a chance to harden.

Allow the first layer of mat laminate to cure. Then add a second layer over it. Allow this to cure.

Turn the laminate over. Using a brush, apply an even coat of resin to the bare side of the core material. Place a layer of dry mat on the resin and press it into the wet resin. Apply additional resin to the mat by again using a dabbing action with the brush. Thoroughly saturate the layer of mat with

resin. When everything looks fine, clean the brush with acetone.

Allow this layer of mat to cure. Ther add a second layer over it. Allow this to cure.

Examine the sandwich core laminate. The ¼-inch core material should have fiberglass skins approximately 3/32 inch thick on each side. The total thickness of the core and fiberglass skins should be about 7/16 inch thick. Although the two skins have a tendency to shear off the core material when certain types of stresses are applied to the laminate, the core laminate has greater structural strength for many applications than a similar four-layer mat laminate without a core. The advantages and disadvantages of core construction will be examined more fully in Chapter 8.

CLOTH AND POLYESTER
RESIN LAMINATE WITH A CORE

This is similar to the previous practice exercise, except two-layer cloth laminates will be added to each side of the core material instead of mat laminates to form a sandwich core molding. Cut out four 6-inch squares of 10-ounce fiberglass cloth.

Place the core material on heavy paper arranged to protect the work area. Pour enough resin for wetting out one layer of cloth into a mixing cup. Add catalyst.

Using a brush, apply an even coat of resin to one side of the core material. Place a layer of dry cloth on the resin, press it down, and smooth it out. Apply additional resin to the cloth with the brush. Thoroughly saturate the cloth with resin. When everything looks good, clean the brush with acetone.

Allow the first layer of cloth laminate to cure. Add a second layer over it. Allow this to cure.

Turn the laminate over. Brush on an even coat of resin to the bare side of the core material. Place a layer of dry cloth on the wet resin, press it down, and smooth it out. Apply additional resin to the cloth until it is thoroughly saturated. When everything looks fine, clean the brush with acetone.

Allow this layer of cloth laminate to cure. Add a second layer over it. Allow this to cure.

Examine the sandwich core laminate. The ¼-

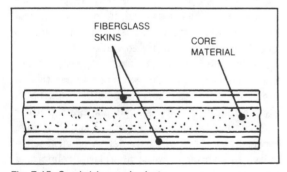

Fig. 7-15. Sandwich core laminate.

inch core material should have fiberglass skins approximately 1/32 inch thick on each side. The total thickness of the core and fiberglass skins should be about 5/16 inch thick.

MAT, CLOTH, AND POLYESTER RESIN LAMINATE WITH A CORE

This is similar to the previous practice exercise, except each skin will have one layer of mat and one layer of cloth. Cut out two 6-inch squares of 1½-ounce mat and two 6-inch squares of 10-ounce fiberglass cloth.

Place the core material on heavy paper arranged to protect the work area. Pour enough resin for wetting out one layer of mat into a mixing cup. Allow this to cure. Then turn the laminate over. Repeat the laminating procedure on this side, adding first a mat layer and allowing it to cure, then finally a cloth layer. Allow time for the laminate to cure.

Add a cloth layer of laminate to the mat layer. Allow this to cure.

Turn the laminate over. Repeat the laminating procedure on this side, adding first a mat layer and allowing it to cure, then finally a cloth layer. Allow time for the laminate to cure.

Examine the sandwich core laminate. The ¼-inch core material should have fiberglass skins approximately 1/16 inch thick on each side. The total thickness of the core and fiberglass skins should be about 3/8 inch.

COMPARISON OF CORE LAMINATES

Compare the three core laminates. The mat lay-up should have the thickest skin. The mat and cloth laminate should be next thickest. The cloth laminate should be the thinnest.

The difficulty of laying up the laminates depends on many factors. All of them should have been fairly easy, though.

While these rigid foam cores add some structural strength to the laminate, the main purpose here is to position the two fiberglass skins apart. Other materials, such as plywood, could also be used as a core material. Depending on the particular lay-up, the plywood core could form a signifi-

cant part of the structural strength of the finished core laminate.

BONDING LAMINATES TO VARIOUS MATERIALS WITH POLYESTER RESIN

Polyester resin gives a poor bond to metals, hardwoods, and other nonporous surfaces. Polyester resin acts as a solvent to some plastic materials such as polystyrene. It cannot be used for bonding a laminate to these plastics. The ability of polyester resin to provide a bond varies from good to poor with other plastics.

Try to gather up scrap pieces of aluminum, steel, a hardwood such as oak, glass, and a variety of plastics. Sand a 1-inch square area on the surface of each material and wash off with acetone. Place these pieces on heavy paper positioned to protect the work area. For each different material, cut a 1-inch square piece of 1½-ounce mat.

Pour enough polyester resin into a mixing cup to wet out all of the mat pieces. The resin can be general purpose, laminating, or finishing.

Catalyze the resin for about 15 minutes of working time, then bond and wet out one piece of mat to each different material. For example, brush on a layer of resin to the bonding area of aluminum. Place a 1-inch square of mat in the wet resin. Press the mat down and smooth it out. Using a brush, saturate the mat with resin. Then repeat on each different material. If the polyester resin happens to soften or dissolve one of the plastics, discontinue the exercise with that material and go on to the next material. When you are finished with the bonding and wetting out, allow the polyester resin laminates to cure.

With a sharp-pointed metal object, test the bond of each laminate. While the results might vary, the bonds were probably poor in most cases. With some of the materials, you might be able to easily peel the laminate off; on other materials the bond might be better. This practice exercise should show that polyester resin bonds poorly to most nonporous surfaces. This same practice exercise will be repeated later with epoxy resin, which generally gives a much better bond to nonporous materials.

POLYESTER PUTTY AND FILLERS

Many polyester resin putty and filling compounds are on the market. These are often formulated for special purposes, such as for marine use or auto body repair. You can also mix up your own polyester putty by adding chopped glass fiber strands and/or milled fibers to polyester resin, either by themselves or with a thixotropic (thickening) powder. For good results, the mixture should contain at least 25 percent by weight of reinforcing material.

As far as fiberglassing repair work is concerned, putty and fillers should be used to supplement laminating work, not as a substitute for it. Use polyester putty and fillers sparingly. Manufactured polyester putty can be used with or without glass fiber reinforcing material.

Several fibrous and powdered additives are used to give polyester resin a putty consistency in manufactured polyester putty and filler compounds. Some of these additives reinforce the resin; others actually weaken it. When using manufactured polyester putty and filler compounds, purchase a small amount and try it out before buying large quantities.

For this practice exercise, you will need a small container of manufactured polyester putty and the necessary catalyst, which is usually sold with the resin putty mixture. The catalyst can be in liquid or paste form, depending on the particular type and brand. Use only the catalyst specifically recommended by the manufacturer. You will also need polyester resin, chopped glass fiber reinforcing material, and thixotropic powder for mixing up your own polyester putty.

First, catalyze a small amount of the manufactured putty. Spread it out in a thin layer on wax paper. Combine chopped glass fibers (these can be pulled from fiberglass mat) and thixotropic powder with polyester finishing resin. Try to get at least 25 percent of the mixture by weight in chopped fibers. Mix well. Then add catalyst. The amount of catalyst to be used depends only on the amount of resin, not on how much chopped glass fiber or thixotropic powder was added. Mix the catalyst using a stirring stick. Spread the putty out in a thin layer on wax paper.

Allow both samples of putty to harden. Compare them first by trying to crack the putty in two using a sharp-pointed metal object. You probably won't be able to do this, at least not easily, with the putty that you mixed. Manufactured polyester putty that I'm familiar with varies from almost as good as the putty you mix yourself to putty that cracks easily. Also note that the putty you mix usually has a hard surface. The manufactured putty might or might not.

Place the two samples on hard surfaces. Tap first one, then the other, with a hammer. The putty that you mixed should hold together by reinforcing material even when the resin is shattered. The manufactured putty might or might not.

For the next practice exercise, you will need one of the mat laminates from a previous practice exercise. Rough sand two areas on the surface for attaching small areas of putty, then wash these areas with a clean cloth soaked with acetone.

Catalyze a small amount of the manufactured putty and apply a thin layer to one of the areas on the mat laminate. Mix up a small amount of your own putty and add catalyst. Apply a thin layer to one of the areas on the mat laminate.

Allow the two areas of putty to cure. Compare them first by trying to pry them loose from the mat laminate with a sharp-pointed metal object. The putty that you mixed probably was bonded well. The manufactured putty might or might not be bonded well.

Tap each sample of putty with a hammer. The putty that you mixed should have held firm. The manufactured putty might have survived the hammer tap, or it might have shattered.

When I require polyester putty for fiberglassing work, I mix up my own as detailed earlier. I have a putty that contains glass fiber reinforcing material. The cured mixture is similar to a mat laminate, though not quite as strong.

When I want the convenience of a manufactured putty, I use epoxy compounds. While epoxy resins are generally much more difficult to apply than polyester resins, this does not apply to some of the newer epoxy putty compounds. Some have fast curing times and are about as easy to apply as most polyester putty compounds. Practice exercises using epoxy putty are given later in this chapter.

SANDING POLYESTER LAMINATES

Practice sanding some of the laminates to see if you can achieve a flat smooth surface. Use a sanding block and aluminum oxide or silicon carbide sandpaper. If there are high spots to be removed, start with a coarse grit sandpaper, but no coarser than necessary to remove the high spots in a reasonable amount of time. Coarse paper also leaves scratches. Gradually work down to finer grits of sandpaper. Use the sanding block.

There might be low areas that require filling with putty. Fill these areas with polyester putty as detailed earlier. A putty knife can be used. Because most polyester putty has some shrinkage during curing, the area should be slightly higher than the desired finished surface. Allow the putty to cure, then resume sanding with a sanding block. Use a fairly coarse grit of sandpaper to remove excess putty, but make certain that the scratches do not extend below the desired finished surface. Some manufactured polyester putty sands easily; others are more difficult to sand. The polyester putty that you mix yourself is generally quite difficult to sand, about the same as sanding a mat and polyester resin laminate.

Finish the sanding with fine grits of sandpaper. Use a sanding block. Do the fine sanding wet or dry.

SUMMARY OF FUNDAMENTAL TECHNIQUES OF FIBERGLASSING WITH POLYESTER RESIN

Most manufactured fiberglass moldings are made with polyester resin. Repair of damage to these fiberglass moldings is usually done with polyester resin.

The techniques covered in the practice exercises using polyester resin form the fundamentals for most fiberglass repair work. If you have mastered these basic skills, you should have little difficulty with the fiberglassing repair work detailed in later chapters. Conversely, if these fundamental skills have not been learned, much difficulty will

probably be encountered in attempting fiberglassing repairs.

I consider the polyester resin portion of this chapter to be the most important part of this book. From this point on, I will assume that you know these fundamental skills and techniques.

FACTS ON EPOXY RESINS

The practice exercises presented in this chapter for *epoxy resins* can be considered optional. They are intended for those people who want to do repair work requiring the use of epoxy resin and for those who want a more complete understanding of fiberglassing.

While polyester resins can be used for most fiberglassing repair work, there are certain applications where it will not work. For these jobs, epoxy resin is often the answer. It has greater bonding strength, better physical characteristics, and less shrinkage than polyester resin. The epoxy resin is usually two or three times more expensive than polyester resin. Epoxy resin has a slower curing time, making it much more difficult to use. Epoxy resins are generally more dangerous to use from a health and safety standpoint than polyester resins.

When using epoxy resins, follow the manufacturer's directions and observe all health and safety precautions. Follow the safety rules and precautions given in Chapter 4. Always wear protective clothing and use proper safety equipment. If any unusual reactions develop from working with epoxy resins, consult a physician.

TYPES OF EPOXY RESINS

Many manufactured epoxy resins are available. Select epoxy resin that is formulated for the particular use which you have in mind. This is usually stated on the container's label.

Epoxy resin requires a matching curing agent or hardener. Depending on the particular formulation, various ratios of resin to curing agent or hardener are used to give the same curing time at a certain working temperature. For example, one type uses half epoxy resin and half curing agent or hardener by volume. The amount of curing agent or hardener used, though, can be reduced by up to 50 percent. Another type is 4 to 1, with 4 parts of epoxy resin being used with 1 part curing agent or hardener by volume. Proportions of from 15 to 25 parts can be used with 100 parts of epoxy resin. Still another type is 6 to 1.

ADDING CURING AGENT OR HARDENER

There are wide variations on the amount of curing agent or hardener to be used. Follow the manufacturer's directions carefully. The measurement can be by volume or by weight, which often differ. For example, one type has a weight of 9 pounds per gallon of resin and 7 pounds per gallon of curing agent or hardener.

After the curing agent or hardener has been added, the time required to harden is generally much shorter when the liquid is in a container or pot than when it is applied in a thin layer to a surface, where the atmosphere cools the surface. To reduce the curing time, heat lamps or other flameless heating devices are often used.

With most types of epoxy resin, there is a range of amounts of curing agent or hardener that can be used. More hardener will reduce the curing time; less hardener will increase it. Unfortunately, changing the proportions often also changes the properties of the cured product.

If you add a pint of curing agent or hardener to a pint of epoxy resin, you get a quart mixture. Remember this when purchasing epoxy resins and hardeners.

The same resin can be used for both laminating the finishing work, so you will only need one type. Achieving a tack-free surface does not depend on inhibiting the air, as is the case with polyester resin.

The slow curing time creates many problems in laminating work. The work is slow because you have to wait so long for the curing to take place. This also makes lay-up work difficult. It is usually impractical to press a piece of reinforcing material into position and hold it there while you wait, say, eight hours for the cure to take place. Use of external heat can greatly speed up the cure, but it's still much slower than when polyester resins are

used. Also, remember that some epoxy resins have faster curing times than others.

Ideally, all the polyester practice exercises would be repeated with epoxy resin. The amount of time that this would take makes it impractical, unless you plan to do a large amount of fiberglassing work using epoxy resins. Also, epoxy resin is expensive, so I have reduced the number of practice exercises. The attempt is to cover the main skills and techniques while keeping the amount of epoxy resin required to a minimum.

Because so many types of epoxy resins are in use, no specific mixing instructions will be given. Follow the manufacturer's directions. The amounts by weight required to wet out a certain piece of reinforcing material will be approximately the same as for polyester resin, as both weigh about 9 pounds per gallon. Because the hardener for epoxy resin might weigh less, often about 7 pounds per gallon, this will throw the weight measurements off a little. Everything should still work out about right when small quantities are being used. Again, remember that when you used polyester resin, consider only the weight of the resin and not that of the catalyst. In the case of epoxy resin, the weight of both the resin and the curing agent or hardener should be added together.

The first practice exercise is to add curing agent or hardener to a small amount of epoxy resin. Using a stirring stick, mix them together. Then pour the mixture out onto wax paper and, using the stick, spread it out into a thin, even layer. Epoxy resins have various viscosities, so some will not pour easily. Use the stirring stick to help get the mixture to the wax paper.

A heat lamp or other heating device can be used to speed up the cure. Follow the manufacture of the epoxy resin's directions in this regard.

Even with external heat of, say, 150 degrees Fahrenheit, it still might require two hours or more for the resin to cure. This is better than the 12 hours or more that it would usually take the same resin to cure at room temperature, especially if you are doing a job that cannot proceed until the resin has cured.

When the epoxy resin has cured, remove it from the wax paper. With a sharp-pointed metal object, try to crack off a piece of the resin. While this can probably be done fairly easily, it is usually more difficult to do than was the case with a similar piece of cured polyester resin.

Place the epoxy resin on a hard surface and tap it with a hammer. It will probably shatter rather easily, but not as easily as a similar piece of cured polyester resin.

ADDING REINFORCING MATERIAL TO EPOXY RESIN

Add chopped strands of fiberglass reinforcing material to the epoxy resin. This can be done before or after the curing agent or hardener has been added. The chopped strands should form about 25 percent of the mixture's weight.

Mix well. Then pour or work the mixture out of the mixing cup to wax paper. Using the stick, spread the mixture out into a thin even layer. Allow the mixture to cure with or without the use of a heat lamp.

After the mixture is hard and no longer tacky on the surface, pick it up from the wax paper. Several things should be noted. The reinforcing fibers add to the mass of the resin. The reinforced resin looks different than resin alone. You can probably see strands of chopped glass fiber on the surface of the mixture. The side that was against the wax paper is probably quite smooth because the wax paper acted like a flat molding or forming surface.

Place the material on a hard surface and give it a tap with a hammer. Try to do this in a similar manner as was done with the hard epoxy resin. Did the reinforced material shatter as easily? Unless something drastic was done wrong, it should be much more difficult to shatter the reinforced resin. Even when it does break up, the reinforcing fibers tend to keep the material from shattering. Although it will probably be difficult to tell the difference from a hammer tap test, it should have been more difficult to shatter the epoxy reinforced material than the polyester reinforced material that was used in an earlier practice exercise.

ADDING EPOXY RESIN TO MAT

Epoxy resin will be added to a 6-inch square piece of 1½-ounce fiberglass mat. The mat should weigh about 0.38 ounce. Mat laminates are typically 25 to 35 percent fiberglass reinforcing material and 65 to 75 percent resin and curing agent by weight. For the 25 percent ratio, about triple the weight of the mat in resin and curing agent is required to fully wet out the mat. For the 6-inch square of 1½-ounce mat, three times 0.38 ounce, or 1.14 ounces of resin and curing agent (combined weight) will be required.

Because some of the resin will probably end up in the brush or otherwise not become part of the finished laminate, I suggest that you mix about 1¼ ounces. This is the combined weight of both the epoxy and the curing agent. Before adding the curing agent to the resin, place the piece of fiberglass mat on a piece of cardboard. Have wax paper ready for transfer of the laminate once one side has been wetted out with resin.

Add the curing agent to the measured out amount of resin. Stir the mixture together. Using a brush, apply resin to one side of the mat, which should be on the piece of cardboard. Use a dabbing action rather than a brushing motion so you do not lump or bunch up the mat.

When one side of the mat is fully saturated with resin, hold the cardboard, turn the mat over, and transfer it to the wax paper so that the wet side of the mat is face down on the wax paper. Put the cardboard aside. Using the brush, dab the remaining resin on the other side of the mat. Saturate the mat and get the resin spread as evenly as possible. When everything looks fine, clean the resin from the brush with epoxy solvent. This must be done before the resin hardens on the brush if the brush is to be used again, but most epoxy resins allow plenty of time for this cleaning.

Allow the laminate to cure with or without the use of a heat lamp. After the laminate has cured to the point where it has a tack-free surface, pick it up from the wax paper and examine it. If everything was done correctly, you should have a mat laminate made up of approximately 25 percent mat by weight and 75 percent resin and hardener by weight. The finished laminate should weigh approximately 1½ ounces. The laminate should be about 1/20 inch thick. The side of the laminate that was placed against the wax paper should be quite smooth, as the wax paper acted as a form or flat mold. The wax paper could have been placed on a curved surface to mold a curved laminate. If everything was done properly and quality materials were used, the laminate should be stronger than a similar laminate laid up with polyester resin.

ADDING EPOXY RESIN TO CLOTH

This is essentially the same as the previous mat exercise, except this time cloth reinforcing material is used. You need a 6-inch square piece of 10-ounce-per-square-yard fiberglass cloth.

Because 1 square yard of 10-ounce cloth weighs approximately 10 ounces, the 6-inch square should weigh about 0.3 ounce, or 1/36 as much. Cloth laminates are typically 45 to 50 percent fiberglass cloth by weight and 50 to 55 percent cloth ratio, the weight of the resin and hardener should be equal to the weight of the cloth. For the 6-inch square piece of cloth, about 0.3 ounce of resin will be required. Because some resin will be lost to the brush and elsewhere during application, mix approximately ½ ounce of epoxy resin and hardener (total weight).

Pour the required amount of epoxy resin into a mixing cup. Before adding the curing agent, place the piece of cloth on a piece of cardboard. After resin is applied to one side, it will be turned onto the wax paper, as was done previously with the mat.

Add the required amount of curing agent or hardener to the epoxy resin. Using a brush, apply resin to one side of the cloth, which should be on the cardboard. Use a brushing action, but take care you do not unravel the unselvaged edges of the cloth.

When one side of the cloth is fully saturated with resin, pick up the cardboard and turn the cloth over onto the wax paper wet side down. Using the brush, work any resin that remains on the cardboard onto the cloth. Put the cardboard aside. Brush the remaining resin onto the cloth. Saturate the cloth with resin and spread the resin as evenly

as possible. When everything looks fine, clean the brush in epoxy solvent.

Allow the laminate to cure with or without the use of a heat lamp or other flameless heating device. After the laminate has cured, pick it up from the wax paper and examine it. If everything was done properly, you should have a cloth laminate made up of approximately 50 percent cloth by weight and 50 percent epoxy resin and hardener by weight. The laminate should be about 1/64 inch thick, much thinner than the previous mat laminate. Compare this to the mat laminate to see if it is so. The mat laminate should be approximately three times as thick as the cloth laminate. The cloth laminate should weigh only slightly more than one-third the weight of the mat laminate.

The side of the cloth that was placed over the wax paper should be relatively smooth, as the wax paper acted as a form or flat mold. Some of the cloth weave pattern probably shows.

LAMINATING TWO LAYERS
OF MAT WITH EPOXY RESIN

An additional layer of mat will be laminated to the previous epoxy mat laminate in this practice exercise. You will need a 6-inch square piece of 1½-ounce mat. Place the mat on the cardboard.

Approximately the same amount of resin and hardener will be required as was used for the first layer of the mat laminate—about 1¼ ounces. Pour the required amount of epoxy resin into a mixing cup. Add the required amount of curing agent or hardener and mix together with a stirring stick. Using a brush, apply resin to one side of the mat, which should be on a piece of cardboard.

Brush on an even layer of resin to the bonding surface of the cured mat laminate. Pick up the cardboard and turn the mat to be added to the laminate over so that the two wet resin areas are face to face. Position the new layer of mat, then smooth it out. Using the brush, dab the remaining resin onto the mat. Saturate the mat and spread the resin as evenly as possible. When everything looks good, clean the resin from the brush with epoxy solvent.

Allow the laminate to cure with or without the use of a heat lamp or other flameless heating device. After the laminate has cured, pick it up and examine it. The second layer of mat should have approximately doubled the thickness of the laminate. With a sharp-pointed metal object, try to separate the two layers at one edge of the laminate. If everything was done properly and quality materials were used, the layers should be bonded together well.

LAMINATING TWO LAYERS
OF CLOTH WITH EPOXY RESIN

An additional layer of cloth will be laminated to the previous epoxy cloth laminate in this practice exercise. You will need a 6-inch square piece of 10-ounce cloth. Place the cloth on cardboard.

Approximately the same amount of resin and hardener will be required as was used for the first layer of the cloth laminate—about ½ ounce. Pour the required amount of epoxy into a mixing cup. Add the required amount of curing agent or hardener and mix together with a stirring stick. Using a brush, apply resin to one side of the mat, which should be on a piece of cardboard.

Brush on an even layer of resin to the bonding surface of the cured cloth laminate. Pick up the cardboard and turn the cloth to be added over so that the two wet resin areas are face to face. Position the new layer of cloth and then smooth it out. Using the brush, apply the remaining resin to the cloth. Saturate the cloth and spread the resin as evenly as possible. When everything looks okay, clean the resin from the brush with epoxy solvent.

Allow the laminate to cure with or without the use of a heat lamp or other flameless heating device. After the laminate has cured, pick it up and examine it. The second layer of cloth should have approximately doubled the thickness of the laminate. With a sharp-pointed metal object, try to separate the two layers of cloth at one edge of the laminate. If everything was done properly and quality materials were used, the layers should be bonded together well.

OTHER METHODS OF
LAMINATING WITH EPOXY RESIN

When adding a second layer to a laminate, an alternate method is to apply resin to the bonding surface of the cured laminate. Place the dry reinforcing material on this, press it down, and smooth it out. Finish by saturating the reinforcing material with resin.

Two or more layers can also be added to a laminate in one operation without waiting for the first layer to cure before adding the second, and so on. This can present problems, as the reinforcing materials might not stay in position until the resin has cured. One possible solution is to place cellophane over the top layer after the laminate has been saturated with resin. Then place a board over this and place a brick or other weight on the board. This works well on flat laminates and can sometimes be adapted to curved ones.

USING EPOXY AND POLYESTER
RESINS IN THE SAME LAMINATE

You can start a laminate with epoxy resin and then finish with polyester resin, or vice versa, provided that the first resin is allowed to thoroughly cure before the second resin is applied. You can use epoxy resin for making crucial bonds and then finish off a laminate with less expensive polyester resin.

As a practice exercise to demonstrate this, add a mat layer using polyester resin to the two-layer cloth laminate that was laid up with epoxy resin. To do this, you will need a 6-inch square of 1½-ounce mat. Pour out approximately 1¼ ounces of polyester resin into a mixing cup and add catalyst. You will need about 7½ minutes of working time. Stir with a mixing stick. Wait about 30 seconds. Brush on an even layer of resin to the bonding surface of the cured epoxy laminate. Place the dry mat on the wet resin. Position it and smooth it out. An alternate method is to wet out one side of the mat placed on a piece of cardboard and then turn it over onto the wet resin on the cured laminate. Using the brush, apply additional resin to the mat. Use a dabbing motion of the brush to avoid bunching up the mat. Thoroughly saturate the mat and spread the resin evenly. When everything looks okay, clean the brush with acetone.

Allow the laminate to cure. Then examine it. Using a sharp-pointed metal object, try to separate the mat layer from the rest of the laminate. If everything was done properly and quality materials were used, the polyester resin should have formed a strong bond with the epoxy resin.

As a practice exercise to demonstrate that this also works in reverse, bond a layer of mat using epoxy resin to one of the polyester cloth laminates from a previous practice exercise. Wash the bonding surface of the polyester laminate with acetone to remove any surface wax that might be present. You will need a 6-inch square of 1½-ounce mat. Approximately 1¼ ounces of epoxy resin and curing agent or hardener (combined weight) will be required to bond and saturate the mat. Pour the required amount of epoxy resin into a mixing cup. Add the required amount of curing agent or hardener and mix together with a stirring stick. Using a brush, apply an even layer of epoxy resin to the bonding surface of the cured polyester laminate. Place the dry mat on the wet resin. Position it and smooth it out. An alternate method is to also wet out one side of the mat with resin before turning it over onto the wet layer of resin on the cured laminate. Apply additional resin to the mat with the brush by using a dabbing action. Thoroughly saturate the mat and spread the resin evenly. When everything looks okay, clean the brush with epoxy solvent.

Allow the laminate to cure. Then examine it. Using a sharp-pointed metal object, try to separate the mat layer from the rest of the laminate. If everything was done properly and quality materials were used, the polyester and epoxy resins should have bonded together well.

OTHER EPOXY RESIN LAMINATES

Many other laminates can also be formed with epoxy resin and fiberglass mat, cloth, and woven roving. Practice exercises for these are not included here because they are not required for most fiberglassing repair work. If you do need to use epoxy resin for a lay-up not included in the pre-

vious practice exercises, you might want to try it first as a practice exercise before using the lay-up on an actual repair job.

ADDING A MAT AND EPOXY RESIN LAMINATE TO WOOD

For this practice exercise, you need a 6-inch square piece of ¼, ⅜, or ½-inch plywood. Rough sand one side of the plywood with coarse sandpaper. This will help give the epoxy resin a better bonding surface. Use cross-grain sanding and scratch up the surface of the wood considerably, while at the same time keeping the surface of the wood fairly even.

You also need a 6-inch square piece of 1½-ounce fiberglass mat. Place the plywood on heavy paper arranged to protect the work area.

Approximately 1½ ounces of epoxy resin and curing agent or hardener (combined weight) will be required to saturate the mat. Because some of the resin will go into the surface of the wood, add an additional ¼ ounce for a total of 1½ ounces.

Pour the required amount of epoxy resin into a mixing cup. Add the required amount of curing agent or hardener to this and mix together with a stirring stick.

Using a brush, apply a layer of resin to the rough sanded surface of the plywood. Place the dry mat in position on the plywood in the wet resin. An alternate method is to place the mat on cardboard and wet out one side of the mat. This is then turned over onto the wet resin on the plywood.

Apply epoxy resin to the mat with the brush. Use a dabbing, rather than brushing, action to avoid bunching the mat up. Once the mat is thoroughly saturated with an even layer of resin and everything looks okay, clean the brush with epoxy solvent.

Allow the resin to cure with or without the use of a heat lamp or other flameless heating device. After the laminate has cured, examine it. Using a sharp-pointed metal object, try to separate the fiberglass laminate from the plywood at one edge. If everything was done properly and quality materials were used, there should be a strong bond—considerably stronger than when polyester resin is used.

ADDING A CLOTH AND
EPOXY RESIN LAMINATE TO WOOD

This practice exercise is essentially the same as the previous exercise, except that cloth will be used instead of mat. You will need a 6-inch square piece of 10-ounce fiberglass cloth.

Rough sand one surface of a 6-inch square piece of plywood. Place the plywood on heavy paper arranged to protect the work area.

Approximately ½ ounce of epoxy resin and curing agent or hardener (combined weight) will be required to bond the cloth to the wood and saturate the cloth. Pour the required amount of epoxy resin into a mixing cup. Add the required amount of curing agent or hardener to this and mix with a stirring stick.

Using a brush, apply a layer of resin to the bonding surface of the plywood. Place the dry cloth in position on the plywood in the wet resin. An alternate method is to place the cloth on cardboard and wet out one side of the cloth. This is then turned over onto the wet resin on the plywood.

Apply resin to the cloth with the brush. Once the cloth is thoroughly saturated with an even layer of resin and everything looks okay, clean the brush with epoxy solvent.

Allow the resin to cure, then examine it. Using a sharp-pointed metal object, try to separate the fiberglass laminate from the plywood at one edge. If everything was done properly, there should be a strong bond—considerably stronger than when polyester resin is used.

The long curing time of epoxy resin can cause problems with air bubbles. It is usually impractical to keep pressing them down and working them out until the resin cures, as is often done when polyester resin is used. One possible solution is to place cellophane over the top layer after the laminate has been saturated with resin. A flat board is then placed over this, with a brick or other weight placed on the board. This often works well on flat laminates and can sometimes be adapted for use on curved laminates.

ADDING POLYESTER
LAYERS TO EPOXY LAMINATES

When covering wood with fiberglass, the first layer can be applied with epoxy resin to give a good bond to the wood. After the epoxy resin has cured, additional layers can be added to the laminate with polyester resin, which is less expensive and easier to work with.

As a practice exercise, add a cloth layer using polyester resin to the plywood and mat and epoxy resin laminate laid up in the previous practice exercise. To do this you need a 6-inch square piece of 10-ounce fiberglass cloth and approximately ½ ounce of polyester resin for bonding and wetting out the cloth. Pour ½ ounce of resin into a mixing cup. Add catalyst and stir. Wait about 30 seconds, then brush on an even layer of resin to the bonding surface of cured epoxy laminate. Place the dry cloth on the wet resin and position it. Smooth out the cloth. An alternate method is to wet out one side of the cloth with the cloth placed on a piece of cardboard. Turn the cloth over onto the wet resin on the cured epoxy laminate.

Using the brush, apply additional resin to the cloth. Thoroughly saturate the cloth with resin and spread the resin evenly. When everything looks okay, clean the brush with acetone.

Allow the laminate to cure. Examine the laminate. Using a sharp-pointed metal object, try to separate the cloth layer from the mat layer of the laminate. If everything was done properly, the polyester resin should have formed a strong bond with the epoxy resin.

Another practice exercise is to add a mat layer using polyester resin to the plywood and cloth and epoxy resin laminate laid up in the previous practice exercise. You need a 6-inch square piece of 1½-ounce fiberglass mat and approximately 1¼ ounces of polyester resin for bonding and wetting out the mat. Pour 1¼ ounces of resin into a mixing cup. Add catalyst and stir. Wait about 30 seconds. Brush on an even layer of resin to the bonding surface of the cured epoxy laminate. Place the dry mat on the wet resin and position it. Smooth out the mat. An alternate method is to wet out one side of the mat and place this wet side down on the wet resin on the cured laminate.

Using the brush, apply additional resin to the mat. Thoroughly saturate the mat with resin and spread the resin evenly. When everything looks okay, clean the brush with acetone.

Allow the laminate to cure. Examine the laminate. Using a sharp-pointed metal object, try to separate the mat layer from the cloth layer. If everything was done properly and quality materials were used, the polyester resin should have formed a strong bond with the epoxy resin.

OTHER METHODS FOR ADDING EPOXY RESIN LAMINATES TO WOOD

Most methods detailed earlier in this chapter for bonding laminates to wood using polyester resin can also be used with epoxy resin. For example, an open weave fiberglass cloth can be applied to the wood dry and fastened in place with staples. The epoxy resin can then be applied to the cloth. This will usually give a good bond to the wood and allow the cloth to be wetted out all in one operation.

Most nonfiberglass reinforcing materials can be applied with epoxy or polyester resin. For example, polyester cloth reinforcing material can be applied to the wood dry and fastened in place with staples. Epoxy resin can then be applied to the cloth. This will usually give a good bond to the wood and allow the cloth to be wetted out all in one operation, as was done in a practice exercise using polyester resin.

While all of the polyester resin practice exercises for laying up core laminates can also be done with epoxy resin, only one practice exercise will be given here. Others from the polyester section of this chapter can be done with epoxy resin for additional practice.

For this practice exercise, you will need a 6-inch square of ¼-inch thick polyurethane, polystyrene or polyvinyl chloride rigid foam. A one-layer mat laminate will be added to each side of core material to form a sandwich core laminate.

Cut out two 6-inch square pieces of 1½-ounce fiberglass mat. Place the core material on heavy paper arranged to protect the work area. The mat laminates will be added one at a time. For the first

one you need about 1¼ ounces of epoxy resin and curing agent or hardener (combine weight).

Pour the required amount of epoxy resin into a mixing cup. Add the required amount of curing agent or hardener and mix together with a stirring stick.

Using a brush, apply an even coat of resin to one side of the core material. Place a piece of dry mat over the resin, press it down, and smooth it out. Apply additional resin to the mat. Use a dabbing action with the brush to avoid bunching up the mat. Thoroughly saturate the layer of mat with resin. When everything looks okay, clean the brush with epoxy solvent.

Allow the resin to cure with or without the use of a heat lamp or other flameless heating device. After the first laminate has cured, turn the laminate over. Apply a mat laminate to this side of the core material in the same manner as was done on the other side. Again, allow the resin to cure.

The ¼-inch core material should have fiberglass skins approximately ¹⁄₂₀ inch thick on each side, forming a sandwich laminate. Although the two skins have a tendency to shear off the core material when certain types of stresses are applied to the laminate, the core laminate generally has greater structural strength for many applications than a similar two-layer mat laminate without a core. The advantages and disadvantages of core construction will be examined in more detail in the next chapter.

The possibility of shaping core materials and then laying up a fiberglass laminate over them also exists. Surfboards are often constructed in this manner. Methods and uses in fiberglassing repair work are detailed in later chapters.

BONDING LAMINATES TO VARIOUS MATERIALS WITH EPOXY RESIN

Unlike polyester resin, epoxy resin usually gives a good bond to metals, hardwoods, and other nonporous surfaces. Epoxy resin will also bond to many types of plastic.

For a practice exercise, gather up scrap pieces of aluminum, steel, a hardwood such as oak, and a variety of plastics. To give good bonding surfaces, rough sand a 1-inch square area or larger section on each different material. Place these on heavy paper positioned to protect the work area. Cut a 1-inch square piece of 1½-ounce fiberglass mat for each material.

Mix enough epoxy resin and curing agent or hardener to bond and wet out all of the pieces of mat. Then bond and wet out one piece of mat to each material. For example, brush on a layer of resin to the bonding area of the aluminum. Position a piece of mat in the resin. Press the mat down and smooth it out. Using a brush, saturate the piece of mat with resin. Repeat on each material. When finished with the bonding and wetting out, allow the epoxy resin to cure.

With a sharp-pointed metal object, test the bond of each laminate. While the results might vary, the bonds to most of all of the materials should be quite good, much better than those achieved in the similar practice exercise with polyester resin. When making fiberglassing repairs on these materials, epoxy resin should be used rather than polyester resin.

EPOXY PUTTY AND FILLER COMPOUNDS

Several epoxy putty and filler compounds are on the market. Many materials, including glass fibers, ground-up steel, stainless steel, aluminum, bronze, lead, and carbide granules, can be added to epoxy resin to form putty and filler compounds. Epoxy putty will bond well to many materials, including steel, iron, aluminum, bronze, brass, wood, ceramic, concrete, plastics, fiberglass, nylon, and glass. Special purpose epoxy putty and filler compounds are available for repairing rubber and making gaskets and other flexible items. While most epoxy putty compounds will not bond to wet surfaces, there is a special epoxy compound that will bond and cure when applied to wet surfaces or even underwater.

You can also mix up your own epoxy putty by adding chopped glass fiber strands to epoxy resin. You can also add thixotropic powder to epoxy resin, with or without adding chopped glass fiber strands, to form a filler compound. Milled glass fibers can be used with or in place of chopped glass fibers.

When purchasing manufactured epoxy putty and filler compounds, select a type that has been formulated for the particular type of job you have in mind. For example, you want a type for marine use that contains no metals that will rust or corrode. For repairing auto bodies, you need a type that has the necessary flexibility and adhesion qualities. Epoxy fillers are available that match the appearance of various metals such as steel, stainless steel, brass, bronze, and aluminum. General-purpose epoxy fillers are available in clear, gray, black, white, and other colors.

For this practice exercise, you need about ½ pint of manufactured general-purpose epoxy putty and the matching curing agent or hardener, which is usually sold along with the epoxy putty. Sometimes the epoxy putty is sold in a kit with the necessary curing agent or hardener, mixing containers, measuring devices, and stirring sticks.

You also need epoxy resin, chopped glass fiber reinforcing material, and thixotropic powder for mixing up your own epoxy putty. First, measure out a small amount of the manufactured epoxy putty. It is important that you don't get any of the epoxy putty left in the container mixed with even a small amount of curing agent or hardener. If you do, the curing will start even if you seal the container back up. Any stick used in the epoxy putty container should not be used in the curing agent or hardener container, or vice versa.

Add the required amount of curing agent or hardener to the epoxy putty. Follow the manufacturer's directions carefully. Some epoxy putty and filler compounds have fast curing cycles even at room temperature.

Mix the epoxy putty and curing agent or hardener together. Use a stirring stick to work it out of the mixing container (which can be a cup, plate, or even a clean piece of wood) and spread it out in a thin layer on a piece of wax paper.

Combine chopped glass fibers (these can be pulled from fiberglass mat) and thixotropic powder to epoxy resin until a thick paste consistency is achieved. Mix these into the epoxy resin with a stirring stick. Add the required amount of curing agent or hardener to a small amount of the paste mixture. Save the rest of the paste mixture for later use.

Mix the curing agent or hardener with the small amount of epoxy putty mixture. Spread the putty out into a thin layer on wax paper.

Allow both putty samples to cure. Compare them first by trying to crack the samples in two using a sharp-pointed metal object. Your ability to crack one or both of them in two depends on many factors, but in most cases, this should not be easy to do.

Place the two samples (or the largest remaining piece of each one) on a hard surface and tap them with a hammer. The manufactured putty might or might not shatter, depending on what type of reinforcing it contained and other factors. The putty that you mixed might have shattered, but it was probably held together to some extent by the glass fiber reinforcing material.

I've found several brands of manufactured epoxy putty compounds that work well for me, so I use these rather than mixing up my own. I also appreciate the fast curing times of some of the manufactured compounds.

For the next practice exercise, you need one polyester laminate from a previous practice exercise and one epoxy laminate. Wash the surface of the polyester laminate to remove any wax that might be present.

Add curing agent or hardener to a small amount of manufactured putty and putty that you mixed up yourself. Apply these in thin layers to small areas of the two laminates. Keep track of which putty is in each location. You should have a sample of each putty on both the polyester laminate and the epoxy laminate.

Allow these to cure. Compare them by first trying to pry the samples loose with a sharp pointed metal object. If everything was done properly and quality materials were used, either the samples could not be pried loose, or it was fairly difficult to pry them off.

Tap each sample of putty with a hammer. The results will vary depending on many factors, but this should give you a general idea of the strength of each sample and allow you to make a rough comparison of the putty compounds.

Try bonding samples of the two types of epoxy putty (the manufactured one and the one you mixed up) to scraps of various materials, such as aluminum, hardwood, steel, concrete, glass, and many plastics. Allow the samples to cure, then try to pry them loose using a sharp pointed metal object. Write the results down in your notebook for future reference. Note which type of putty worked best with each material and which materials allowed satisfactory bonding.

SANDING EPOXY LAMINATES

Practice sanding some of the epoxy laminates from the practice exercises. Use the same techniques as were used for sanding polyester laminates, as detailed previously in this chapter.

SUMMARY OF EPOXY PRACTICE EXERCISES

While many types of fiberglassing repair work can be done with the less expensive polyester resin, there are certain jobs that call for epoxy resin. The epoxy practice exercises were designed not only to give practice in handling and applying epoxy resin but to show some of the advantages of using epoxy resin instead of polyester resin for certain types of work.

BASIC CONTACT MOLDING WITH POLYESTER RESIN

Contact molding techniques can be applied to both repair and construction projects, as detailed in later chapters. The purpose here is to introduce the basic techniques for contact molding.

In contact molding, one side of the molding is formed against the mold surface. Shallow rectangular shapes are introduced here; more complex shapes and deeper molds are detailed in later chapters.

Shallow-Depth Rectangular Contact Molds

Shallow-depth rectangular contact molds can be formed from a variety of materials, such as from fine-grained wood using ordinary woodworking techniques. Because contact molding gives a smooth surface on one side only, it is important to note that there are two basic forms a shallow-depth rectangular contact mold can take. If you want to mold a shallow rectangular tray with a smooth upper surface, the contact mold surface will look like the bottom surface of a tray (Fig. 7-16). If you want to mold a shallow rectangular tray with a smooth under surface, the contact mold surface will look like the upper surface of a tray (Fig. 7-17). To mold a shallow rectangular tray with a smooth upper surface, the mold is placed with the contact surface upward (see Fig. 7-16). The molding is then laid up over this (Fig. 7-18) and removed from the mold after it has cured (Fig. 7-19). If it is to be used as a tray, it is then turned over so that the smooth molded surface faces upward.

Contrast this to a mold for a shallow rectangular tray form with a smooth under surface (see Fig. 7-17). The mold is positioned like an upright tray, and the molding is laid up inside (Fig. 7-20) and removed from the mold after it has been cured (Fig. 7-21). Depending how the molding is to be used, it can be positioned like a tray with a smooth underside or a cap or lid with a smooth top. In each case, remember that the molding is not a duplicate of the mold but a reverse image.

Shallow rectangular molds can be shaped from a single piece of wood, or two or more pieces can be glued together. Ordinary woodworking techniques are used for shaping shallow and medium-depth rectangular molds from wood. Figure 7-22 shows a wood mold suitable for molding a shallow tray form with a smooth upper surface (when the

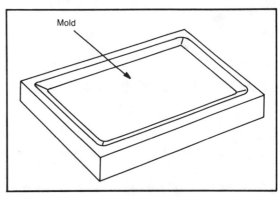

Fig. 7-17. Concave contact mold gives molding with smooth convex side.

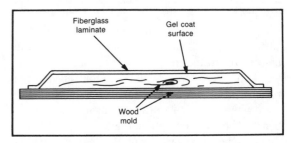

Fig. 7-18. Laying up molding over a convex mold form.

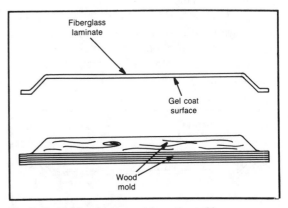

Fig. 7-19. Molding is removed from the mold.

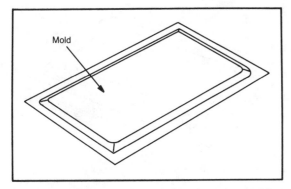

Fig. 7-16. Convex contact mold for tray form gives molding with smooth concave side.

tray is turned upright). Figure 7-23 shows a wood mold suitable for molding a shallow tray form with a smooth lower surface (when the tray is turned upright).

89

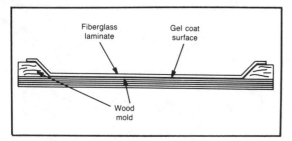

Fig. 7-20. Laying up molding over a concave mold form.

After the wood has been shaped and sanded, the next step is to prepare the wood surface so that it is suitable for molding. One method is to coat the molding surface with a thin layer of polyester resin. In addition to the shaped and sanded wood mold, you need either finishing or general-purpose polyester resin, catalyst for the resin, a mixing cup, a small mixing stick, one longer mixing stick that can be used to stir the resin in the container in which it was purchased, and a small brush for applying the resin. You also need heavy paper to protect the area where you are working, which can be on a bench, floor, or concrete surface outside. While a mold can be placed directly on the floor, it might be more convenient to place it on a raised surface. You should also wear proper protective clothing and safety equipment.

Position the wood mold with the molding surface upward. When you have everything ready, begin by opening the container that the resin came in. Carefully stir it with a clean mixing stick. When removing the stick from the container, allow as

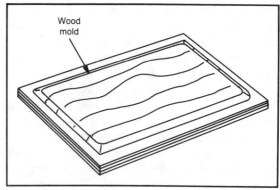

Fig. 7-22. Wood mold for laminating tray shape with smooth concave side.

much of the resin to drain off the stick back into the container as possible. Use a clean rag to clean off the remaining resin from the stick and put the stick aside for later use. Use this one stick only for stirring uncatalyzed resin in the containers in which it is sold.

You need about 1 ounce of resin for each square foot of wood surface to be covered. Pour the required amount of resin into the mixing cup. Replace the lid on the resin container and set it aside.

If available, add about 1/8 ounce of styrene monomer to the resin and mix thoroughly. While the resin can be applied to the wood surface without this, the thinning action of the styrene thins the

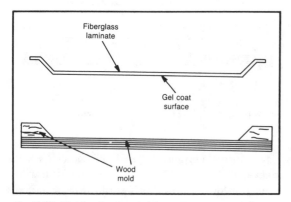

Fig. 7-21. Molding is removed from the mold.

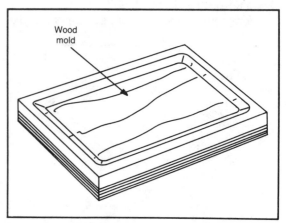

Fig. 7-23. Wood mold for contact molding tray form with smooth convex side.

90

resin and allows it to better penetrate the surface of the wood.

Next, add 8 drops of catalyst for each ounce of resin to the resin. This is approximately 2 percent by volume, which should give about 15 minutes of working time. Wait about 30 seconds, then brush a smooth thin layer of the catalyzed resin onto the molding surface (Fig. 7-24). When you have finished applying the coat of resin to the molding surface of the wood and everything looks okay, clean the brush with acetone before the resin hardens on the brush. Allow the resin to harden on the wood surface. Even though this should happen about 15 minutes after the resin is applied, it is best to wait at least an hour before attempting to sand the surface.

Next, lightly sand the surface with fine-grit paper to remove the surfacing agent. This is not necessary if laminating resin is used. An alternate method is to remove the surfacing agent by wiping the surface with acetone on a clean cloth.

Then apply a second coat of polyester resin in the same manner as detailed above. When this has cured, apply a third coat in the same manner. When the third coat has cured, sand with fine-grit wet/dry sandpaper and water.

The next step is to polish the surface using a buffing compound. This can be done by hand with a clean cloth or with a power buffer.

It is important to take the necessary time to make a good mold. Any defects in the mold surface will be copied on the moldings made with the mold.

A variety of existing shallow-depth rectangular forms can be used as molds. Shallow depth trays and pans made of stainless steel and aluminum are ideal. Many plastic trays and pans also work, though the plastic must be a type that polyester resin will not dissolve. Keep in mind that you are not making a duplicate of the tray or pan but are using it as a mold to obtain a reverse image. For contact molding, you will use one side of the tray or pan when laying up a molding, although many trays

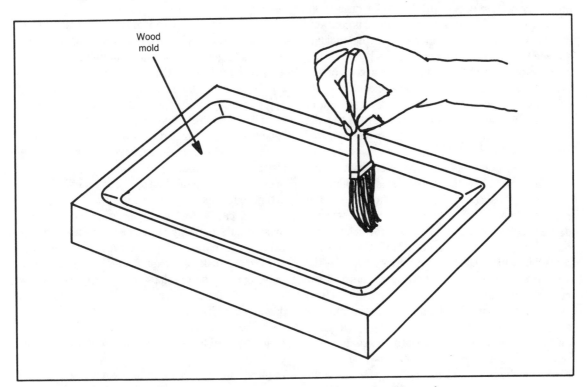

Wood mold

Fig. 7-24. Brush a smooth thin layer of catalyzed polyester resin on the wood molding surface.

and pans will allow molding on either side. Make certain that the mold form does not have vertical sides or overhangs that will make removing a finished molding difficult or impossible.

Shallow and medium-depth rectangular contact molds can also be made from a variety of other materials. For contact molding, you will use only one side of the mold, either for a concave or convex molding form.

Contact Molding Shallow and Medium-Depth Rectangular Forms

To contact mold a shallow-depth rectangular form, you need a suitable mold, a paste-type mold release, polyester gel coat resin, laminating and/or finishing or general-purpose polyester resin, catalyst for the resin, mixing cups, short and long mixing sticks, a small brush for applying the resin, and fiberglass reinforcing material (mat is recommended for first attempts at molding).

The shallow-depth rectangular mold is placed on the protective paper with the mold surface upward. If the molding surface of the mold is concave, do the lay-up inside a pan form. If the molding surface of the mold is convex, do the lay-up over a caplike form. In either case, apply a coat of the paste-type mold release using a clean cloth to spread a thin layer of the release over the entire mold surface. The purpose of this is to keep the molding from sticking so that it can be removed after it has cured.

The next step is to apply the gel coat on the mold surfaces over the mold release. When you have everything ready, open the gel coat polyester resin container. This can be clear resin or any desired color, or color pigments formulated for polyester gel coat resin can be added. Carefully stir the resin with a clean mixing stick. Remove the stick from the container, allowing as much of the resin to drain into the container as possible. Use a clean rag to clean off the remaining resin from the stick and put the stick aside for later use. Save this one stick for stirring only uncatalyzed gel coat resin in the containers in which it is sold.

Pour the required amount — depending on the area to be covered, about 1 ounce per square foot

— of the gel coat resin into a mixing cup. Replace the lid on the gel coat resin container and set it aside.

Next, add the catalyst to the resin. This is approximately 2 percent by volume, which is usually used for polyester gel coat resin at 75 degrees Fahrenheit. This can vary for the particular brand of gel coat resin being used, however. Follow the manufacturer's directions. Wait about 30 seconds, then brush a smooth thin layer of the catalyzed gel coat resin onto the mold surface over the mold release agent. Use a continuous motion with the brush (Fig. 7-25) rather than painting back and forth. The gel coat should have a thickness of only .02 to .03 inch. A thicker layer will be brittle and subject to cracking and crazing. When you have finished applying the coat of resin and everything looks okay, clean the brush with acetone. Allow the resin to harden.

Gel coat resin usually doesn't contain a wax additive. The laminate can therefore be laid up directly over this with no additional surface preparation.

The next step is to laminate the first layer of fiberglass reinforcing material, which is usually mat, to the gel coat. Cut the reinforcing material to a size slightly larger than is required to cover the molding area of the mold. Beginners are advised to apply one layer of reinforcing material at a time and allow this to set before applying the next layer. As experience is gained, two or more layers (up to a maximum of about 6 ounces of reinforcing material per square foot) can be applied wet. This is usually more difficult than applying a single layer at a time but makes the work faster.

Mat can be formed to shallow curves without making cuts but only after the resin has been applied to the material. Sharp curves and angles usually require making cuts and overlapping the mat in these areas. The overlaps can be worked down to an even thickness after resin has been applied by dabbing with the brush.

The next step is to calculate the amount of resin required for saturating the reinforcing material that is to be applied wet at this time. Pour the required amount of resin into a mixing cup. You will probably need about 7½ minutes of working

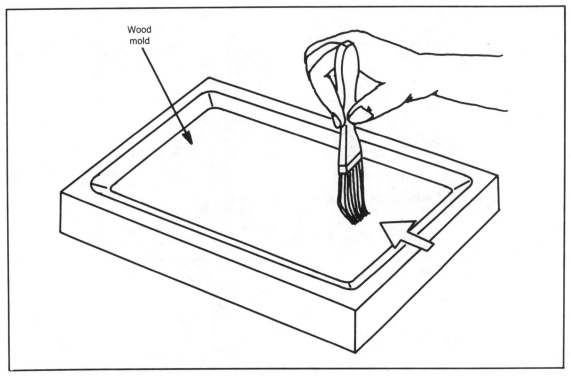

Fig. 7-25. Apply gel coat to the mold surface with long continuous strokes of brush.

time to apply the resin to a single layer of reinforcing material; more time might be required for multiple layers.

Add the required amount of catalyst to the resin and stir. Wait about 30 seconds. Then, using the brush, apply a coat of resin over the gel coat layer that was applied to the mold surface and allowed to set previously.

Position the first layer of fiberglass reinforcing material (usually a mat layer) over the wet resin so it is centered, press it down, and smooth it out. For mat, use a dabbing action rather than a brushing action to avoid lumping up the mat.

Saturate the reinforcing material with resin (Fig. 7-26). Spread the resin to saturate as evenly as possible, including areas of the reinforcing material that extend slightly beyond the desired size of the finished shallow-depth rectangular form. These can be trimmed off later after the laminate has cured.

If a second layer of reinforcing material is to be applied, wet over the first layer, position it in the wet resin, press it down, and smooth it out using the brush. If the second layer is cloth rather than mat, use a brushing action but be careful not to unravel the cut edges of the cloth.

Saturate the second layer of reinforcing material with resin. Spread the resin as evenly as possible. Saturate all areas of the reinforcing material.

Apply all layers of reinforcing material that are to be applied wet at this time. When everything looks okay, clean the resin from the brush with acetone.

It generally works best if laminating resin is used for all but the final layer of resin on the backside of the laminate. The final layer should be sanding or finishing resin to give a surface that is not sticky so that it can be sanded.

Allow the laminate to cure. If additional layer or layers of reinforcing material are to be added to the laminate, measure out and catalyze the required amount of polyester resin. Brush on a layer of wet resin over the cured laminate. Position the first additional layer of fiberglass reinforcing

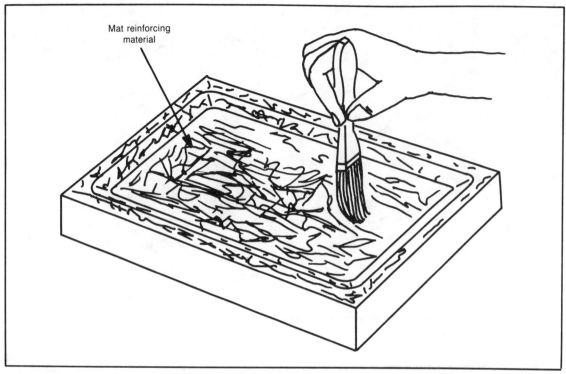

Mat reinforcing
material

Fig. 7-26. Saturate the reinforcing material with resin.

material over the wet resin. Center it, press it down, and smooth it out. The brush can be used for this.

Saturate the reinforcing material with resin. Spread the resin as evenly as possible. If additional layers are to be applied wet, these should be added at this time in a similar manner.

Allow the laminate to thoroughly cure, then remove it from the mold. Even though the mold release was used, it still might be necessary to pry the laminate loose with a putty knife or other similar tool.

This completes the chemical part of the construction. For most projects, the fiberglass molding requires trimming. First mark the pattern and saw off the excess fiberglass at the lip of the tray or pan form. Sawing methods are detailed in Chapter 6. Next, file and sand the edges smooth. Polish and buff the gel coat surfaces and, for some projects, the edges of the laminate.

Chapter 8

Molding Methods Used in Manufacturing

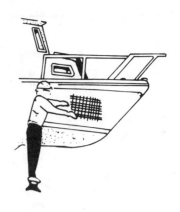

An understanding of the molding methods used in manufacturing fiberglass products is helpful for fiberglassing repair work. The fundamentals of fiberglassing were covered first in Chapter 7, so this material will be more meaningful. There are two basic types of molding used in manufacturing: *contact* and *pressure,* with many variations of each method. Most fiberglassing repair work involves some form of contact molding, although pressure molding techniques can sometimes be employed.

CONTACT MOLDING

Contact molding is a fiberglass lay-up in or over a male or female mold or form without pressure being applied to the side of the laminate away from the mold or form surface — that is, other than the contact pressure. Simple forms of this method were introduced in Chapter 7. For example, a sheet of wax paper was used as a mold or form for a flat surface. The resins and fiberglass reinforcing materials were laid up over it. While air bubbles were worked out, it was basically the weight of the resins and reinforcing materials that held the mixture against the wax paper until it cured.

Contact molding is done by *hand lay-up,* with the resins being applied to the mold surface by brush, rollers, and squeegees; by *spray up,* with the resins and reinforcing material being sprayed from a chopper gun; or by some combination of the two methods. For either method, the gel coat resin is usually sprayed onto the surface of the mold.

Advantages and Disadvantages

One advantage of contact molding over pressure molding is that molds and other equipment are generally less expensive. Also, relatively inexperienced workers can do hand lay-up work. Another advantage is that, with contact molding, large fiberglass structures can be molded that would be impractical to do using pressure molding.

A main disadvantage of contact molding is that the back side of the laminate will not be as smooth and fair as the front side because there is no way to mold or form the back side during the cure of the laminate. Other disadvantages are that contact molding is slower, involves hand labor, and is somewhat less accurate when small parts with complicated shapes are required.

Sanding

When contact molding is used in manufacturing, the side of the laminate that is to be smooth and finished almost always goes against the mold surface. The mold for a bathtub (Fig. 8-1) or swimming pool is almost the reverse of that used for a boat hull (Fig. 8-2) or car body. In each case the desired finished side is the side that will be against the surface of the mold and is applied first in the laminating process. This is usually the case regardless of whether a resin gel coat is to be applied inside or over the mold as the first layer of the laminate, or the laminate is to be painted later. Manufacturers want to avoid sanding, which in-

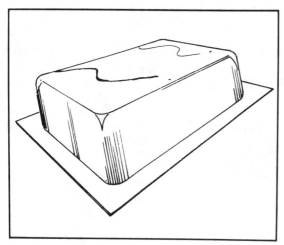

Fig. 8-1. Contact mold for a bathtub.

Manufacturers have developed methods for avoiding sanding and fairing of large areas of fiberglass when contact molding is used. For example, in boat manufacturing, finished gel coat surfaces are frequently needed both outside and inside a hull. To accomplish this by contact molding, two separate moldings are used, one for the outside of the hull and a hull liner that goes inside the hull (Fig. 8-3). The two moldings are then bonded together to form a boat that has smooth gel coat surfaces on both the outside and inside.

Some one-off construction methods do involve sanding and fairing. These can be done over a low-cost male mold or form, thus avoiding the expensive cost of tooling a female mold. Plenty of sanding and fairing is required, and it is extremely difficult to achieve a smooth and fair surface comparable to that possible from a quality female mold.

volves plenty of labor even when power sanding equipment is used. Imagine the sanding that would be necessary if the bathtub or swimming pool were molded in reverse. The rough side would be the side that needs to be smooth. Even if all the sanding necessary to achieve this were done, it would be difficult to obtain the surface possible from against a mold. Each molding would be slightly different in the finished side from the previous one. Also, the gel coat layer would have to be sprayed on last, like paint, instead of against the mold first. This would not matter if the molding was to be painted rather than gel coated.

Male and Female Molds

There is considerable confusion in the way the terms "male" and "female" molds are used. Usually, if the surface of the laminate that is to be the "outside" finished surface goes on first against the mold, it is considered as a female mold. The boat hull mold shown in Fig. 8-2 is easy to recognize as a female mold, but this is not the case with the "female" mold for the bathtub shown in Fig. 8-1, which by appearance looks like a plug or male mold.

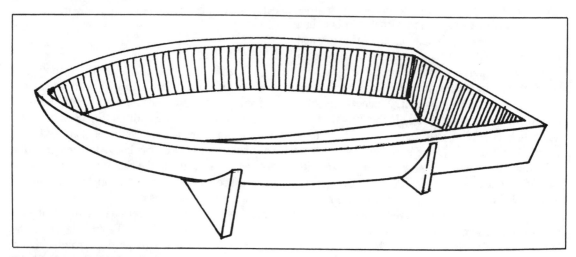

Fig. 8-2. Contact mold for a boat hull.

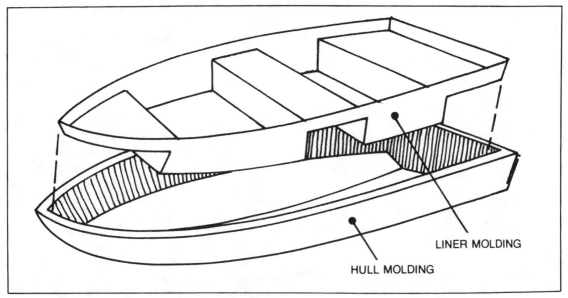

Fig. 8-3. Use of separate liner molding in boat hull molding.

Most manufacturing using contact molding is done with female molds. As mentioned, it costs a lot of money to make a quality female mold. Once the mold is made, though, hundreds of fairly accurate duplicates of the same molding can be made from it. The duplicates must be done one at a time. The mold is used for the lay-up of one laminate or molding, which is then removed from the mold. The same mold is then used to lay up a second molding. While there are some limits to the useful life of a mold—they do eventually wear and/or become damaged—it is not unusual for 100 or even 1,000 moldings to be made from a single mold.

Before a female mold is constructed, the usual procedure is to have a design made up (either on paper or as a small scale model) of what the finished molding is to look like. This is not the only way. Many fiberglass molds have been taken from existing car bodies, boats, swimming pools, bathtubs, and so on, including from those made out of materials other than fiberglass. There are sometimes problems involved here. The design for a specific thing made out of wood, metal, or concrete might not be the best design for fiberglass construction. Still, successful molds have been "designed" in this manner.

The trend today is to design especially for the material that is to be used for construction. The design aspects of fiberglass are covered later in this book. The main ideas, however, are to make the best use of the material, to make it structurally sound, and to make it reasonably easy to mold.

Making a Plug

A *plug* is constructed to the design lines in the shape of the finished surface desired for the fiberglass molding. A plug for a bathtub is shown in Fig. 8-4. Notice that it looks like a bathtub. A plug for a boat hull is shown in Fig. 8-5. Notice that it looks

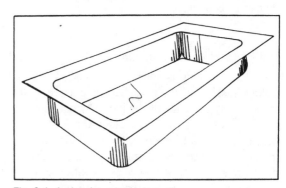

Fig. 8-4. A plug for a bathtub mold.

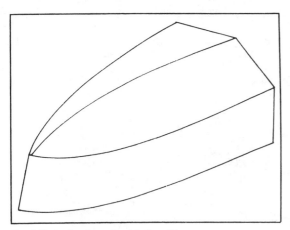

Fig. 8-5. A plug for a boat hull mold.

like a boat hull. The idea is to duplicate these shapes.

The plugs can be made from wood, plaster of paris, clay, concrete, various plastics, or a combination of materials. The plug must be as close as possible to the design lines or desired shape of the finished molding, as a mold will be taken from the plug to duplicate its shape. Any unevenness or lack of fairness in the plug will be transferred to the mold and, in turn, to fiberglass laminates made in the mold. While it is possible to make some changes in the mold itself after it has been made, this is extremely difficult. It is much easier to get the plug exactly the way it should be, so changes will not have to be attempted on the mold itself. On the plug, you are working with the shape desired for the finished laminate. On the mold, you are working with the shape surrounding the laminate that is to be molded.

Making a Mold from a Plug

Once the plug is finished, the next step is to make a mold from the plug. This might seem like the long way to go about the job. Why not skip the plug and just construct the mold in the first place? This can and has been done, but it is very difficult to make a mold that is accurate, smooth, and fair in this manner. You would be working with the shape of the area surrounding the desired object, and

with complicated shapes, this involves a difficult change in thinking.

Molds can be made from several materials, but a fiberglass lay-up seems to work best for most uses. Essentially, a fiberglass laminate is laid up over the plug. Remember, however, that the term "plug" is being used to mean shapes like a plug (the boat hull, for example), flat surfaces, and pans (the bathtub, for example).

First, the plug is sprayed or otherwise coated with a wax releasing agent or *polyvinyl alcohol (PVA)* parting agent. The plug can thus be removed from the fiberglass mold once the lay-up has been completed, and it has cured. While methods vary, a typical one is to first wax the plug, buff this out, and then spray on a layer of PVA.

A special color gel coat, called *tooling gel coat*, is sprayed on. A color is selected that will contrast sharply with the gel coat color of the items to be molded later. Tooling gel coat is a special tough gel coat formulated for use on molds.

The tooling gel coat is allowed to harden. A layer of fiberglass mat usually follows. Polyester resin is used in most cases. The mat layer is used first to give the mold a smooth surface. The weave pattern of cloth and especially woven roving will frequently show through on the surface.

The laminate is then followed by layers of desired reinforcing material. Frequently, layers of mat are alternated with woven roving to make the mold stiff and strong. The mold must be constructed so that it will not flex or sag when a laminate is laid up in it. Additional reinforcing is frequently required. This can be from materials like metal and wood. The reinforcing often acts as a base for wheels, so the mold can be moved about the work area, and as a frame so the mold can be turned to desired angles for lay-up work.

The mold can be in one piece if there are no undercuts, such as the one shown in Fig. 8-6, which would not allow removal of a laminate from the mold. The mold could be removed from the plug by destroying the plug, but this still would not solve the problem of later removal of a molding from the mold.

If undercuts are present, the mold can be made in two or more pieces, which are clamped or

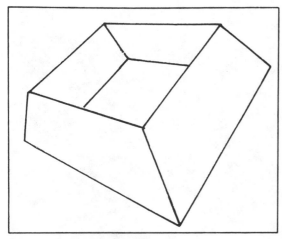

Fig. 8-6. Mold with undercuts.

bolted together later when molding work is being done. The mold can then be taken apart to remove the molding.

Another reason for making the mold in more than one piece is to avoid difficult molding jobs, such as deep keel areas on boat hulls, which are difficult to lay up. The mold is made in two or more pieces. The pieces are later molded separately and then bonded or otherwise fastened together (Fig. 8-7).

Because attaching separate moldings together can cause structural problems, one-piece moldings are used when practical (Figs. 8-8 and 8-9). Assembly of separate moldings is covered in later chapters.

If the mold is to be in more than one piece,

Fig. 8-7. Boat hull mold in two sections. The hull is molded in halves that are later bonded together.

Fig. 8-8. Mold allows molding in near level plane.

Fig. 8-9. Moldings are joined together to form the boat shell.

this must be taken into account in the tooling and lay-up of the mold. Once the lay-up of the mold is completed, the resin is allowed to cure. Reinforcements, if required, are sometimes added before the plug is removed. In other cases the plug is removed first, then the reinforcing members are added. These are attached to the mold by fiberglass bonding or other means (see Chapter 9).

Preparing the Mold for the Lay-up of the First Laminate

When the mold is completed, it is prepared for the lay-up or molding of the first laminate. It is generally easiest to do fiberglass lay-up work on a fairly level surface. Inclined and vertical surfaces are more difficult. Overhead surfaces are even more difficult and generally avoided. The mold is usually positioned so that most of the lay-up work is done from directly above (Fig. 8-8). For a bathtub, no single position of the mold is satisfactory. In manufacturing, this difficult lay-up work is either tolerated, or the mold is arranged so it can be turned to various angles. The mold can be angled so that the area where the lay-up work is being done is as near level or horizontal as possible. This is a fairly simple matter in the case of a small mold, but it is much more difficult for large objects such as swimming pools and boat hulls. Special hoists and other equipment are frequently used for changing the angle of the mold.

To get ready for a lay-up, the mold is first cleaned. Wax is then applied and buffed out. A parting agent like PVA is usually sprayed on, especially for the first few laminates. The wax alone might then be enough to allow removal of moldings from the mold.

Lay-up of Laminate Layers

The first layer of the laminate is usually a sprayed-on layer of gel coat in the desired color. A recent trend is to omit the gel coat on some moldings and then to paint the molding later after it has been taken from the mold. The reasons for this are problems with the gel coat (cracking, crazing, and color fading) and difficulty of application. Also, gel coats are frequently damaged when the molding is

taken from the mold, which requires difficult touch-up work. Also, paints for use on fiberglass have improved greatly over the years. Using two-part polyurethane paint, it is possible to get a gel coat-like appearance that is in many ways more durable than traditional gel coat applications.

The gel coat, if used, will be the finished side of the laminate. The gel coat is allowed to harden. Because an air-inhibited resin is usually used, the surface on the exposed side will remain tacky.

The next layer of the laminate is fiberglass mat, although a lightweight cloth with a smooth finish is sometimes used. A thin layer of resin is applied over the gel coat. The mat is placed over the wet resin. Additional resin is applied to the mat until it is saturated with resin.

The remaining layers of the laminate are added over the first layer of mat. The laminate should be designed to give the desired characteristics to the molding for its particular application. The laminate can be all mat layers. Cloth and/or woven roving can be alternated with mat layers.

Considerable care in the lay-up work is needed to produce quality laminates. The ratio of resin to reinforcing material, which varies depending on the reinforcing material used, is kept as close to the ideal ratio as possible. Avoid resin-rich (too much resin) and resin-starved (too little resin) areas.

Instead of laying up the laminate by hand, sometimes a *chopper gun* is used. This is a spraying device that mixes catalyst and resin and chops up strands of fiberglass, spraying the mixture onto the mold. This is applied over the gel coat layer until the desired thickness of laminate is achieved —all in one operation. This method is much faster than the hand lay-up, but a poorer quality laminate results. Still, for many applications it is adequate. It takes a highly skilled operator to spray up a laminate of even thickness or to desired varying thicknesses without getting too much resin for the amount of chopped glass fiber reinforcing strands or, less frequently, too little resin.

Moldings are also made by combinations of hand lay-up and spray-up with the chopper gun. For example, the chopper gun can be used to replace layers of mat, with fiberglass cloth and woven roving laid up by hand over wet layers of the spray-up. The cloth and woven roving are usually saturated with resin by brush or roller application. Allow for curing before adding the next spray-up layer.

Large Moldings

Large moldings present special problems. Many pieces of reinforcing material have to be used. When joining pieces, they are usually over-lapped from about 1 to 6 inches or more, depending on the particular application. These overlaps must not all be made in the same areas if an even thickness is desired. The overlaps are made at random in most cases. In some moldings, the cloth and woven roving reinforcing materials run continuously the length of the molding in one direction, or other specific patterns are used to give desired structural strength to the laminate.

The thickness of the laminate is often varied, to give the greatest strength in the areas where it is required. This presents additional difficulties in the lay-up of the laminate. The idea is to put the materials where they will do the most good. These and other structural concepts are treated in more detail in later chapters.

On a large laminate, one additional layer of reinforcing material can add greatly to the labor and material cost of the lay-up. Follow the designed lay-up as closely as possible. Laminates are de-signed to be as strong as necessary with, hopefully, a healthy safety margin. With the present high material and labor costs, the trend is to reduce the safety margin. This means making the best possible use of materials.

Most large moldings require a number of days for the lay-up work to be completed. One layer of reinforcing is saturated with resin at a time and is allowed to harden before the next layer is added. This makes it easier to remove air bubbles and wrinkles in the lay-up. Sometimes two and even three layers are added on top of each other before any have set. This is not a job for the inexperienced, however.

Most lay-up work in manufacturing is done during 8-hour workdays, often with no lay-up work being done on weekends. This requires careful lo-gistics to keep track of the lay-up work that has been done and what needs to be done next. There is usually a laminating schedule. Each job is checked off as soon as it is completed.

Layers of laminates are occasionally left out of moldings. This is sometimes accidental. In other cases it is done to reduce the laminate's cost. Sometimes unsafe products result. One boat that sunk because the fiberglass hull split was raised, and the damage was analyzed. The hull had only about half the advertised thickness in the area of damage, and many of the layers of reinforcing material had been left out. There was a lawsuit over this, though I never did learn if the hull was deliberately or accidentally built to the wrong specifications.

Most large moldings require additional rein-forcement in the form of beams, stringers, ribs, frames, hat sections, and so on. This results in much more efficient use of materials than to at-tempt to achieve the same result by increasing the laminate's thickness. Structural concepts such as this are covered in greater detail in later chapters. Remember that the molding, as it comes from the mold, might require additional attachments. The same applies to assembly. Moldings must be joined together in many cases. Making moldings is usually only part of the fiberglass manufacturing process.

When the lay-up of a molding has been com-pleted, the laminate is allowed to cure for a certain length of time, usually from several hours to sev-eral days depending on the type and size of the laminate, before it is removed from the mold. Most manufacturers want to remove the molding from the mold as soon as possible, so the mold can be used to start another molding. Moldings can be damaged or ruined if they are taken from the mold too soon.

Removing Moldings from Molds

Even though the molds are waxed and PVA is applied to the mold before the lay-up, it is often difficult to remove moldings, especially large ones, from the molds. Hoists are often used for "break-ing" the molding from the mold. For example, a boat hull might have plywood bulkheads bonded in place while the hull is still in the mold. The hoist is connected to these bulkheads, and the hull is worked free from the mold by hoisting.

Once the molding is removed from the mold, the wax and polyvinyl alcohol (PVA) are cleaned from the gel coat (Fig. 8-10). Touch-up work is usually required on the gel coat, as it is difficult to achieve perfection in the original molding, especially if it is a large one.

Some moldings will be rejected for one reason or another. Others will require repairs to make them satisfactory for use. Various manufacturers have different standards here. Some moldings from the same mold come out better than others.

Core Constructions with Double Skins and Other Moldings

The previous description is for a *single-skin* contact molding. In many cases core constructions are used, often with *double skins.* Typical core materials include honeycomb cell paper, foamed plastics, and wood, especially balsa and plywood (Fig. 8-11). The first skin or laminate is laid up in the mold as described. The core material is then placed over a wet layer (saturated with resin) of reinforcing material or otherwise bonded to the laminate. A second skin is then laminated over the core material. This is usually all done before the laminate is removed from the mold.

Sometimes other structural reinforcements are also added to the laminate before it is removed from the mold. The structure of laminate must be such that it will retain its shape when removed from the mold. Care must also be taken to properly support the molding if it is taken from the mold before fully cured or before structural reinforcing has been added.

Sometimes the molding is the finished prod-

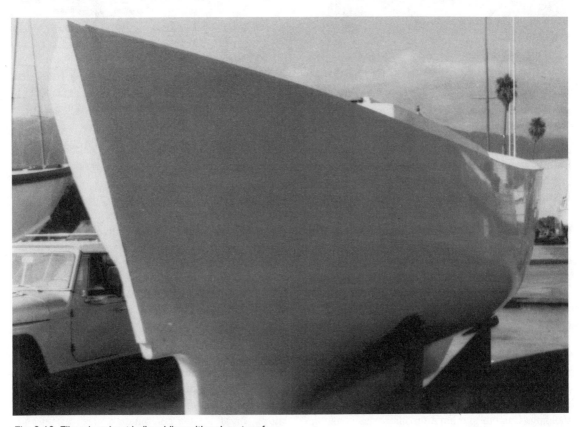

Fig. 8-10. Fiberglass boat hull molding with gel coat surface.

Fig. 8-11. Fiberglass boat deck has a plywood core.

uct—the way it will be marketed and sold. Bathtubs and hot tubs are examples. Other moldings require additional construction and assembly. In some cases the moldings are only a small part of the total construction. For example, a travel trailer might have only the roof made of fiberglass, with the rest of the construction of wood, metal, and other materials.

Some moldings are laid up upside down, such as tubs and swimming pools; others are right side up, such as boat hulls. Special equipment is required for turning over large items like swimming pools. Sometimes this is done with the item still in the mold. In other cases it is removed from the mold upside down and then turned over.

Many fiberglass products that will require repair have laminates that were made by contact molding. Custom and amateur fiberglass construc-

tions also employ contact molding, although less expensive male molds are usually used.

Quality of Contact Moldings

You must be able to recognize contact moldings and have some idea of the quality of the laminates in order to make effective repairs to them. Fiberglass moldings that were formed by contact moldings can be recognized because the back side is not as smooth and even as the side that was against the mold, which is usually gel coated or painted.

Often it will be possible to tell if the contact molding was hand laid up or sprayed up with a chopper gun. Examine the back side of the molding. Overlapping of reinforcing material and/or the weave of cloth or woven roving indicates hand lay-up, or at least a combination of spray-up and hand lay-up. A spray-up laminate will show chopped strands of fiberglass with a matlike appearance, except there will be no overlapping of pieces of mat reinforcing material.

Woven roving on the back side of the laminate indicates not only hand lay-up, but also that heavy reinforcing materials were used. This can be important in situations where a strong laminate is required.

Other clues to the quality of contact moldings include:

☐ **Thickness.** How does the thickness of the product compare to those made by other manufacturers? While a thicker laminate does not necessarily mean a stronger one, it often does.

☐ **Fairness.** If possible, sight along the finished side of the molding. It should be smooth and even. Lack of fairness could indicate a poor mold was used, or the molding was not properly reinforced.

☐ **Ability to Support Loads.** When you put weight on the laminate, how much does it give? The amount acceptable depends on how the laminate is to be used, but the laminate should be able to support all loads normally placed on it without undue flexing.

☐ **Gel Coat.** The gel coat often gives a good indication of the quality of the entire laminate. Cracking, crazing, and fading of gel coat indicates poor quality of materials and/or poor application. This often indicates lack of quality in the remainder of the laminate. A good gel coat, in turn, indicates the use of quality materials and proper application of them. This is especially true if the laminate has been in use for a long time.

PRESSURE MOLDING

In an attempt to avoid some of the problems associated with contact molding, manufacturers use various methods of pressure molding, such as *matched die, pressure bag,* and *vacuum bag.* While only limited use of these methods is made in fiberglassing repair work, an understanding of the processes can be useful in making repairs on them.

Matched Die Molding

Matched die molding is a pressure method of molding a fiberglass laminate by squeezing catalyzed resin and fiberglass reinforcing materials between two molds (Fig. 8-12). This method is now commonly used for certain types of fiberglass molding. It gives a quality laminate that is smooth and formed on both sides. A proper ratio of resin to glass fiber reinforcement can be maintained. Production is much more rapid than when contact molding is used, especially when the molds are heated to speed up the curing time.

Disadvantages are that it is expensive to tool for matched die molding, and this method is generally limited to fairly small moldings. For example, to date, boat hulls only up to about 20 feet in length have been produced on a regular basis by this method. Experiments have been carried out to determine the feasibility of making even larger hulls and other moldings by this method.

Chevrolet *Corvette* car bodies are made by matched die molding. The first Corvette car bodies were produced by contact molding. Then vacuum bag molding (described later in this chapter) was used to produce several hundred car bodies before going on to matched die molding, which allows volume production.

Matched die molding does not allow application of a thin color gel coat to the laminate. The Chevrolet Corvette car bodies are painted after they are removed from the matched die molds. It is possible to add a color pigment to all the resin used. The color resin will then be throughout the laminate. This method has only limited usefulness, however, as the glass fibers show through the surface.

Matched die molding is practical only when a large number of the same molding is to be produced. The tooling is very expensive and must be amortized by making many moldings.

Simple matched die molds can be constructed. Mat and catalyzed resin is applied to one surface of the mold by hand. The matched die mold is be set in place and clamped down in the desired position. Most production equipment does some or all of these jobs automatically.

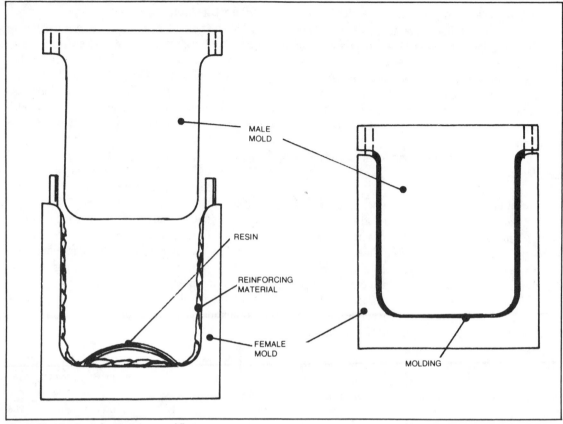

Fig. 8-12. Matched die fiberglass molding.

Modern matched die molding allows rapid molding cycles. A fiberglass sports car top having an area of about 20 square feet has a molding cycle of about five minutes in matched die molding, whereas it commonly takes several hours or more using hand lay-up in a contact mold. Before going to matched die molding, most manufacturers first use molding methods that have less tooling expense to make certain that there is enough demand for the product to justify the high tooling cost required for matched die molding.

Pressure Bag Molding

Pressure bag molding uses one rigid die with the opposite die, usually the male, made of a flexible rubberlike material (Fig. 8-13). The required amount of catalyzed resin and reinforcing material is fed into the mold. Air pressure is forced into the

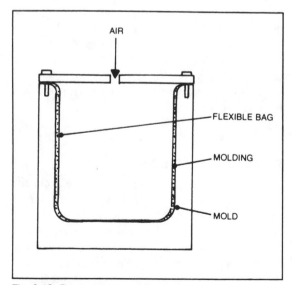

Fig. 8-13. Pressure bag molding.

area above the bag. This squeezes the resin and reinforcing material into the desired position, with a uniform surface against the rigid die and a nearly uniform surface against the bag. The pressure is maintained until the molded part has cured.

To speed the cure time, hot air or steam pressure can be used. The mold is heated in some other manner. Using this method, items such as helmets can be molded in three-minute cycles.

Vacuum Bag Molding

Vacuum bag molding is similar to the pressure bag method except that a vacuum is used between the rigid die and the bag (Fig. 8-14). The area between the bag and the rigid mold must be airtight for this method to work. Molding is done by first adding the required amount of reinforcing material and catalyzed resin to the mold. This is done automatically on modern systems. Vacuum is then applied. The flexible bag is drawn down into the desired position and the resin and reinforcing material are squeezed into shape. This produces a uniform surface against the rigid die and a nearly uniform surface against the bag. The vacuum is maintained until the molded part has cured. Various means of applying external heat to the mold are used to speed the cure. Items such as helmets can be molded in three-minute cycles.

OTHER FIBERGLASS MOLDING METHODS

More refined and automated molding techniques are constantly being tried. Limited applications have already been made using *continuous laminating,* a process where resin and reinforcing materials are added to each other, shaped into desired form, and processed or cured in a continuous manner. Flexible *plunger moldings* have also been used. Even better methods are likely to be developed and used in the future.

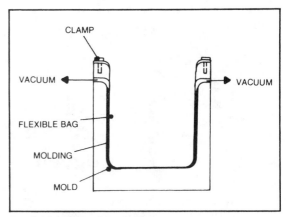

Fig. 8-14. Vacuum bag molding.

QUALITY OF FIBERGLASS MOLDINGS

The quality of fiberglass moldings for a particular use depends on the molding methods used, the soundness of design of the molding, quality of molds and equipment, quality of materials used, and care and skill in fabrication. Fiberglass molding, from a manufacturing point of view, is a highly competitive business. A first requirement is a marketable fiberglass product. This must then be produced at a cost so that it can be marketed profitably. Superior design and quality of construction can increase production cost, which must be made up for by a higher selling price.

Molding costs can be reduced by many ways, some of which result in a lower quality molding. The quality of original moldings has a considerable bearing on the ease with which repairs can be made if the molding is damaged.

If you are interested in doing fiberglassing repair work, you should learn as much about the manufacturing of fiberglass moldings as possible. One way to do this is to visit fiberglass manufacturing plants and observe the various molding methods used. By understanding how the moldings were made, you should be better prepared to repair them when they are later damaged.

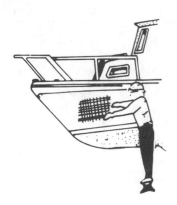

Chapter 9

Structural and Assembly Concepts and Practices

A basic understanding of the structural and assembly concepts and practices used in fiberglass construction is helpful for doing fiberglassing repair work. Fiberglass moldings, whether of single-skin or sandwich core construction, are usually only part of a finished product. In most cases additional structural reinforcement must be added to the moldings, and fiberglass moldings must be joined together by bonding and/or mechanical fastening. Often the moldings are joined with other materials. Examples are a fiberglass car body to a metal chassis and a fiberglass boat shell to wooden interior components. Various fittings are frequently attached to fiberglass — faucets and drains to fiberglass bathtubs and cleats to fiberglass boat decks.

SINGLE-SKIN FIBERGLASS LAMINATES

Two basic types of fiberglass laminates are commonly used: *single skin* and *sandwich core*. Most fiberglass laminates are single skin. These can be molded in a variety of ways (see Chapter 8). The laminates are made up of two basic components — fiberglass reinforcing material and resin. The fiberglass reinforcing material in a laminate provides the strength, and the resin provides a means of holding the glass fibers in position.

The strength of fiberglass to this point has been considered only in a very general way. There are different types of strength that are important in fiberglass laminates, including tensile strength, flexural strength, compressive strength, and shear

strength. As compared to other materials that can be used for similar types of construction, fiberglass laminates rate well in some types of strength and not so well in others.

Because of the great variability in fiberglass laminates, due to factors such as care and skill used in fabrication, quality of materials, type and arrangement of reinforcing materials, thickness of laminate, and so on, large variations can exist from one laminate to another. This makes strength testing of fiberglass laminates quite complex.

Tensile Strength

Tensile strength can be defined as the resistance of a material to a force tending to pull it apart. This is often measured in pounds per square inch. Fiberglass laminates generally have high tensile strength in directions parallel to the layers of the laminate (Fig. 9-1). The tensile strength of mat and polyester resin laminates is often about 18,000 to 20,000 pounds per square inch. This is generally much higher than most types of wood, which average about 7,000 pounds per square inch with the grain and only about 600 pounds across the grain. It is about the same as one type of aluminum, although there are other types of aluminum that have tensile strengths of up to about 44,000 pounds per square inch. It is less than steel. One type of steel has a tensile strength of 30,000 pounds per square inch, and structural steel has about 60,000 pounds per square inch.

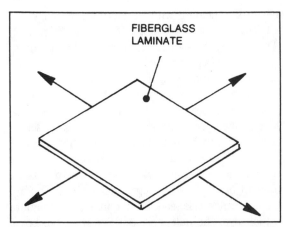

Fig. 9-1. Fiberglass laminates generally have high tensile strength in directions parallel to the layers of the laminate.

A cubic foot of each of these materials has a different weight. The fiberglass mat and polyester resin laminate weighs about 92 pounds, wood averages about 42 pounds, aluminum about 176 pounds, and steel about 490 pounds.

Woven roving laminates generally have higher tensile strength than mat laminates. Cloth laminates are usually higher in tensile strength than woven roving laminates. The cloth laminates frequently have tensile strengths of 40,000 pounds per square inch or more.

The tensile strength is actually more related to the volume of glass fiber in the laminate than the type (mat, cloth, or woven roving) of reinforcing material used. Mat laminates have the lowest tensile strength because they absorb the most resin when wetted out. Woven roving uses less resin and has a higher tensile strength. Cloth uses even less, giving a laminate with the highest tensile strength.

It might appear that mat would have the lowest tensile strength because of the short strands of glass fiber and that woven roving and cloth would be higher because the strands run continuously the length and width of the material. Tests indicate that this is not the case. There is little loss of tensile strength until the strands become less than ¾ inch long.

Cloth and woven roving laminates have the greatest tensile strength in the two directions of the weave. They have relatively less tensile

strength on the diagonal and all other directions in the plane of the laminate. The directions of the weaves of laminate layers can be varied to give fairly uniform tensile strengths in all directions of the plane of the laminate. Mat has the lowest tensile strength because of the high resin content, but it has the most uniform tensile strength in the plane of the laminate because of the glass strands' randomness.

The tensile strength perpendicular to the layers of the laminate is much lower than the tensile strength in the plane of the laminate in laminates laid up in layers. This is mainly because loadings perpendicular to the laminate tend to pull the laminate apart in the relatively weak resin areas (Fig. 9-2). This must be considered when assembling and attaching moldings together and attaching things to them.

Some types of molding (see Chapter 8) are done in one operation rather than in layers. These usually use some form of mat or chopped strands. Laminates produced by these methods are often considered to be *isotropic*. Their physical properties are essentially the same in all directions, like aluminum and steel. These laminates have about equal tensile strength in all directions.

Laminates that are laid up in layers, including mat laminates, are *orthotropic*. Their physical properties vary with direction.

Fiberglass laminates also have high *flexural strength* (in some cases greater than structural steel), *compression strength,* and *impact strength.* The *shear strength* varies from quite good to poor,

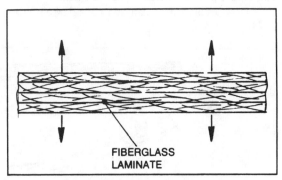

Fig. 9-2. Loadings perpendicular to layers of laminate tend to pull laminate apart in relatively weak resin areas.

depending on the type of laminate and method of molding or lay-up.

Flexibility of a Laminate

The main design and construction problem associated with fiberglass laminates is that fiberglass laminates are very flexible, about six times more flexible than aluminum that has about the same strength, and about 20 times more flexible than steel. Notice that flexibility and strength are two different factors. A steel wire is strong and flexible. An egg shell lacks flexibility or is rigid and weak.

Fiberglass laminates have flexibility similar to plywood; they are springy but not rubbery. Fiberglass laminates are also stronger and more costly than plywood. This means that in most constructions they must be thinner. Extra support and stiffening will usually be required, as it is generally too expensive to accomplish this by increasing the thickness of the laminates unnecessarily.

Stiffening a Laminate

Several stiffening methods are used. Some of these methods are involved in the design and construction of single-skin laminates. Others are covered in later sections of this chapter.

Stiffening of a single-skin laminate can be accomplished by the shape of the molding. Curvatures and corrugations, for example, have greater stiffness than flat panels of the same thickness. These are usually part of the designing process but can sometimes be added to molds later.

Use of curvatures can add considerable stiffness to a molding with no increase in weight or cost over that of molding a flat panel (Fig. 9-3). This is ideal as fiberglass can be molded in curvatures, including compound curves, about as easily as it can be molded in flat sections.

Some of the first boats constructed from fiberglass were actually wood designs. The molds were often taken from existing wooden boats. These frequently had hard chimes and large flat areas on the hulls. When constructed of fiberglass, they required unnecessarily thick laminates and/or a large amount of added stiffening by means of stringers, frames, bulkheads, and other reinforcing members attached to the hull shell. As time went on, boats were designed especially for fiberglass construction. Curved sections were used instead of flat sections and sharp angles, which are difficult to mold. The hulls had greater stiffness for the same laminate thickness, and these required less stiffening by means of added reinforcing members.

Another method of stiffening a laminate is to mold in corrugations, such as the ones shown in Fig. 9-4. This is sometimes done on the bottom of boat hulls that have large, relatively flat areas. If these were not included in the original mold and a need for them is discovered later, they can sometimes be added by modifying the mold. Usually this is quite difficult, and the options are quite limited. It is far better to add them in the design stage, before the molds are constructed. Early fiberglass constructions were usually much stronger than necessary, and this also involved the use of more material. Sometimes the material was not used wisely. The structures were stronger than necessary in some areas, but too weak in others. Much

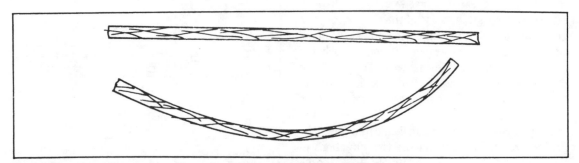

Fig. 9-3. Curved panel has greater stiffness than a flat panel of the same thickness.

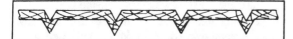

Fig. 9-4. Molded-in corrugations add stiffness.

Fig. 9-6. Gradual change in laminate thickness.

more is known about fiberglass design and construction today, with the result that better use of the material can be made.

Another method of adding stiffness is to increase the thickness of the laminate selectively — that is, to vary the thickness of the laminate. With some materials, this is difficult to do. For example, in the case of plywood planking, usually one thickness of plywood is used throughout. Most pressure molding techniques allow almost unlimited variations in thickness of moldings. In the case of contact moldings, additional layers can be added to the laminate where more thickness is desired.

This is also a way to repair an existing fiberglass molding that does not have sufficient stiffness. Techniques for doing this are covered in later chapters.

When varying the thickness of laminates, avoid abrupt changes in thickness. Abrupt changes in thickness, such as shown in Fig. 9-5, cause a stress concentration at that point. Changes in thickness should be gradual over a distance to produce an even deflection rather than a large stress concentration at one point, as is the case when abrupt changes in thickness are used. In the case of hand lay-up in a contact mold, in the areas where the greatest thickness is desired, progressively larger or smaller pieces of reinforcing material are used to give a gradual change in thickness (Fig. 9-6).

Another method of stiffening a single-skin

laminate is to mold in *flanges*. This is especially useful for stiffening the edges of laminates (Fig. 9-7). While a similar effect can usually be achieved by bonding and/or mechanically fastening a piece of wood or other material to the edge of the molding, molding in flanges often makes this unnecessary (Fig. 9-8).

Another method of stiffening the edge of a laminate is to add extra thickness at the edge of the laminate (Fig. 9-9). To avoid large stress concentration at one point, the thickness change should be gradual.

SANDWICH CORE CONSTRUCTION

Another way to add stiffness to a laminate is to use *sandwich core construction*. Basically, the laminate is divided into two skins, which are positioned apart by adding a core material between them. To be effective, the core must form a link between the two skins. This forms a much stiffer

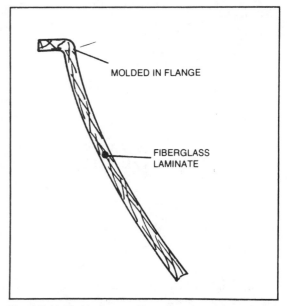

Fig. 9-7. Molded-in flange stiffens the edge of the molding.

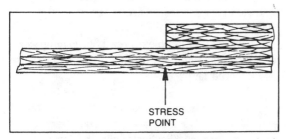

Fig. 9-5. Abrupt change in the thickness of fiberglass laminate causes a stress concentration at that point.

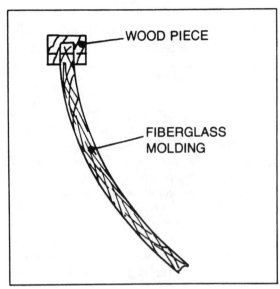

Fig. 9-8. Stiffening edge of fiberglass molding with a wood piece.

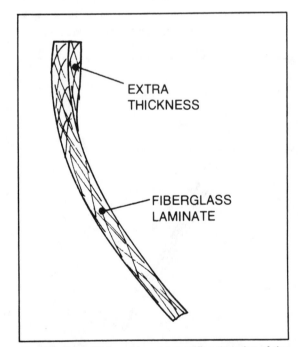

Fig. 9-9. Extra thickness is used to stiffen the edge of the laminate.

laminate than if the two skins are side-by-side in a single laminate. Several materials are used for cores, including wood, plywood, rigid plastic foams, and honeycombs made from paper, fiberglass, and other materials.

Sometimes the cores are added during the lay-up. For example, in a contact mold the outer skin of the laminate is first laid up in the mold. The core material is then bonded to this, often by setting it in a wet layer of resin. The inner skin is then bonded and laminated over the core material.

The two skins are molded separately in some cases. The core is then bonded and/or mechanically fastened in place between the two skins.

The cores add varying degrees of structural strength to the laminate in addition to spacing the two skins apart. For the sandwich construction to be effective, there has to be at least a certain amount of structural support provided by the core material. The design aspects of core construction are quite complex, but generally the core must be strong enough to hold the two skins in position relative to each other. The core material must have adequate structural strength and be bonded well or otherwise attached to the fiberglass skins, so the sandwich core laminate acts as a single structural unit.

There are several problems associated with core construction. One of these is low shear strength, especially when using lightweight core materials that have little structural strength. This problem can generally be surmounted by proper design and selection of core material.

Another problem is in attaching things to the laminate, which can cause the two skins to pull apart from the core or squeeze together and crush the core material. Methods for attaching things to sandwich core laminates are covered later in this chapter.

Still another problem area is in joining sandwich moldings together. Methods for doing this are also covered later.

Core materials add not only stiffness to a laminate but also insulation and, in some cases, buoyancy floatation. These are important factors in some applications like boat hulls and superstructures. In some cases cores have absorbed water, but this can usually be prevented by proper design, selection of core material, and construction.

Plywood is frequently used as a core material for stiffening and/or reinforcing boat decks, cabin tops, and transoms. Plywood can be laid up in a laminate in the same way as when rigid plastic foam cores are used.

Most of the strength and structural concepts described earlier in this chapter for single-skin laminates also apply to sandwich core laminates, when the sandwich core laminate is considered as a single unit.

Sandwich core laminates can present repair problems when they are damaged. For example, if there is a hole all the way through the laminate, the core and both skins require repair, like doing three jobs instead of one. As will be detailed in later chapters, sandwich core laminates can usually be repaired when they are damaged.

METHODS OF STIFFENING LAMINATES

For both single-skin and sandwich core moldings, additional stiffening can still be required. Commonly used methods include adding stiffeners molded from fiberglass, braces, frames, ribs, bulkheads, hat sections, other reinforcing members, and molded fiberglass liners.

Molded Fiberglass Stiffeners

Figure 9-10 shows a molded fiberglass stiffener for a boat hull. This is bonded inside the hull laminate either before or after the hull molding is removed from the mold. When properly designed and constructed, a molded fiberglass stiffener can be more effective than if the same amount of resin and reinforcing material, as used for molding the stiffener, was used to increase the thickness of the hull molding.

Molded fiberglass stiffeners of various designs, shapes, and sizes are also used in other fiberglass constructions, including car bodies and recreational vehicle components. Molded fiberglass stiffeners can serve more than one purpose. For example, a molded fiberglass unit for a boat might be used not only as a stiffener for the hull shell but also as a base for mounting an engine (Fig. 9-11).

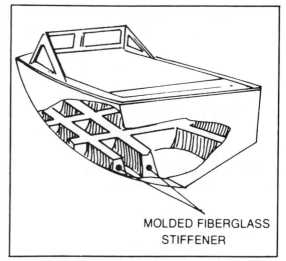

MOLDED FIBERGLASS
STIFFENER

Fig. 9-10. Molded fiberglass stiffener for boat hull.

Molded fiberglass stiffeners can be attached to fiberglass moldings by bonding and/or mechanical fastenings. In the areas of attachment, large stress concentrations must be avoided if the molding that is being stiffened and supported is to retain its original shape. The areas of contact between the stiffeners and the laminate should be large. Abrupt changes in thickness should be avoided. The contact areas should gradually taper down to the thickness of the laminate being stiffened (Fig. 9-12). If the contact areas are small and/or the changes in thickness are not gradual, high stress areas, frequently called "hard spots," can occur. These can cause distortion of the laminate and, if

Fig. 9-11. Molded fiberglass engine bed installed in molded fiberglass boat hull.

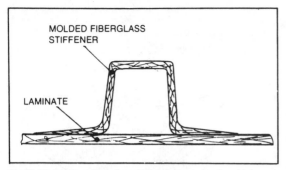

Fig. 9-12. Molded fiberglass stiffener tapers to the thickness of the laminate being stiffened.

the laminate flexes or "works" during use, even laminate failure.

Special problems can occur when molded fiberglass stiffeners are mechanically fastened to laminates. The fasteners must be properly spaced so that high stress areas are not formed. The use of mechanical fasteners for making attachments to fiberglass moldings is detailed later in this chapter.

Laminating Fiberglass Stiffeners Over Core Materials

Several stiffeners can be molded in place over core materials. The core material might or might not add strength to the molded stiffener. In some cases the core material acts merely as a form or male mold for laminating the fiberglass stiffener. In other cases the core material also adds structural strength to the stiffener.

One popular type is a molded-in-place rib or hat section (Fig. 9-13). Core materials include cardboard, rigid plastic foam, plastic pipe, and wood. This type of stiffener can be added during the original construction before or after the molding is removed from the mold, or at some later time if it is found that the molding requires additional stiffening. Methods for laminating these stiffeners in place are detailed in later chapters.

These stiffeners should contact a large area of the laminate being stiffened to spread the load over as wide an area as possible and taper down gradually to the thickness of the laminate. Design for size, shape, construction, and arrangement of the stiffeners is important for achieving maximum stiffening with a minimum amount of materials and added weight.

The core materials used for laminating these in place range from nonstructural to structural. Special care must be taken when using structural cores such as wood, which have different physical properties than fiberglass (Fig. 9-14). For example, if heavy wood beams are used as cores for laminating fiberglass stiffeners in place, it might be necessary to reinforce the laminate in the area where

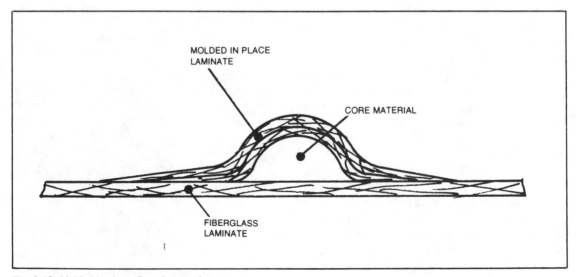

Fig. 9-13. Molded-in-place rib or hat section.

114

Fig. 9-14. Wood beams serve as structural core for stiffeners on boat hull.

the beam will be positioned. Add extra layers to the laminate and/or use a rigid plastic foam pad between the laminate and the wood.

Several shapes of stiffeners are used, including square and rectangular, triangular, and half round. Patterns such as webs can be used (Fig. 9-15). These should be designed to give maximum stiffening and support to the laminate, while at the same time being cost, material, and labor-efficient. It is also important to keep weight to a minimum for some applications.

Laminating stiffeners over core materials usually allows better bonding to the laminate being reinforced than the previously described method of first molding a stiffener from fiberglass and then bonding it in place. Laminating in place allows the stiffener to be filled with a core material instead of air space. Molded stiffeners can also contain core materials. For example, rigid plastic foam can be used to fill in spaces between laminate and molded stiffener. Pour or spray-in-place plastic foam can also be used. This is applied in liquid form and gives a rigid plastic foam when cured.

Laminating Other Types of Stiffeners in Place

Fiberglass stiffeners are sometimes laminated in place using removable forms and molds, which can then be used over again. Figure 9-16 shows a typical angle mold. This can be clamped or taped to the laminate. A fiberglass angle is then laminated in place. A release agent is used so that the angle laminate will not stick to the angle mold, and the angle mold can be removed.

A second angle laminate can then be laid up against the first one to form a T *section* (Fig. 9-17). A double angle or Z *section* can be laminated in place using a removable mold or form in a similar manner (Fig. 9-18). When contact molding is used, these stiffeners can be added before or after the molding has been taken from the mold.

Laminating Edge Stiffeners

One method of stiffening and strengthening the edge of a laminate is to add *beading*. Wire,

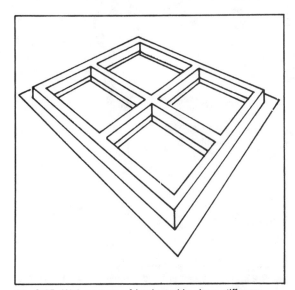

Fig. 9-15. Web pattern of laminated-in-place stiffeners.

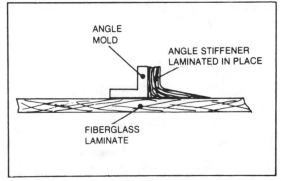

Fig. 9-16. Removable angle mold for laminated angle stiffener in place.

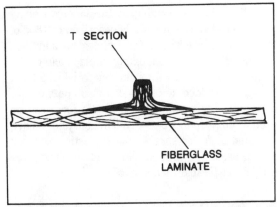

Fig. 9-17. Second angle laminate laid up against the first one to form a T section.

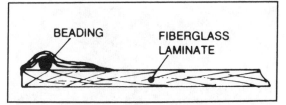

Fig. 9-19. Beading is used to stiffen the edge of the laminate.

Braces, Frames, Beams, and Ribs

Braces, frames, beams, and *ribs* are used to stiffen, strengthen, and support fiberglass laminates. These can be molded from fiberglass in a female mold and then bonded or otherwise attached to the laminate. They can also be molded in place. Still another method is to construct the braces, frames, beams, or ribs from wood, metal, or other materials, then bond and/or mechanically fasten them to the laminate.

Notice that these support and reinforcing members are added to the molding after it has been

cord, rope, plastic rod, and other things can be used as a core or form to laminate the beading over (Fig. 9-19).

Another means of stiffening and strengthening the edge of a laminate is to add a *bulbous cap laminate* to the edge (Fig. 9-20). Split cardboard tubing, plastic pipe, and other items can be used as a core or form for laminating the fiberglass bulbous cap to the edge of the laminate. This method not only stiffens the edge of the laminate but keeps the edge of the laminate from delaminating.

The edges of laminates are particularly weak and vulnerable areas. They are problem areas in the design and fabrication of many fiberglass products.

Beading can be added inside a contact molding before or after the molding is removed from the mold. Bulbous caps must usually be laminated in place after the molding has been removed from the mold, as the mold would be in the way.

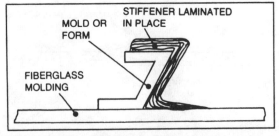

Fig. 9-18. A Z section stiffener laminated in place using a removable mold or form.

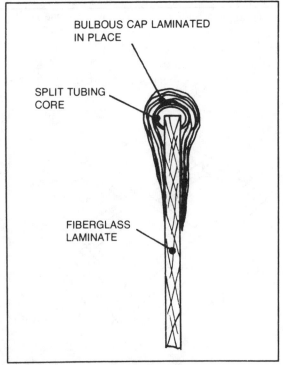

Fig. 9-20. Bulbous cap is used to stiffen the edge of the laminate.

molded, either before or after the molding has been removed from the mold. This is the reverse order of many types of construction. For example, the frames and supports for a roof are constructed first, then the roof is added over them.

Braces, frames, beams, and ribs need to be designed so that they effectively stiffen, strengthen, and support the fiberglass laminate without creating hard spots and other high-stress areas or points that could deform or damage the laminate. In some cases the appearance of these structural members is also important; in other cases they will be hidden from view, such as structural members that are underneath a fiberglass swimming pool that will be set in the ground. When wood, metal, and other materials besides fiberglass are used, the physical properties of the materials (expansion, contraction, and so on) need to be considered. The present manufacturing trend is toward using fiberglass structural members for stiffening, strengthening, and reinforcing fiberglass moldings. The structural members will have physical characteristics similar to those of the fiberglass molding. Fiberglass also bonds well to fiberglass. Bonding other materials to fiberglass can sometimes be difficult.

Plywood Bulkheads

Plywood bulkheads are often used in fiberglass boats (Fig. 9-21). Similar structural members are used in other fiberglass products. These are frequently bonded to fiberglass moldings with angle strips of fiberglass that are laminated in place. If the plywood bulkhead is placed directly against the hull molding or other molding, a high-stress area or hard spot can be created on the fiberglass laminate. The hard spot is due to the slight expansion and contraction of the fiberglass laminate with temperature changes, almost no expansion and contraction of the plywood, and from working and twisting of the laminate, such as that of a boat hull when in use.

To prevent this, a space is frequently left between the edge of the plywood and the laminate. A pad of rigid plastic foam or other material can be used between the plywood and the fiberglass laminate (Fig. 9-22).

Fig. 9-21. A plywood bulkhead is fiberglass bonded to molded fiberglass boat hull.

The fiberglass angle bonding strips should contact a fairly wide area of the fiberglass laminate and gradually taper to the laminate. This can be accomplished by laminating the angle bonding strips with progressively wider pieces of reinforcing material (Fig. 9-23).

Sharp angles between the plywood and the laminate should be avoided with the bonding laminate, so the strain will be taken up by rounded angles (Fig. 9-24). A rigid plastic foam pad is often shaped to give these cures.

Molded Fiberglass Liners

Most manufacturers try to avoid most of the previously described methods of adding stiffeners to fiberglass moldings, especially if they will show

117

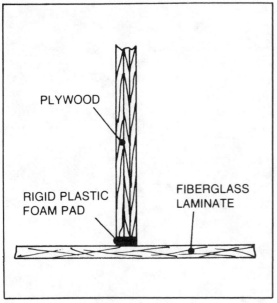

Fig. 9-22. Rigid plastic foam pad used between plywood and fiberglass laminate.

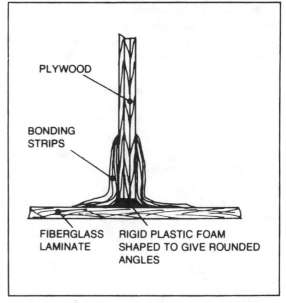

Fig. 9-24. Rigid plastic foam shaped to give rounded angles.

in the finished product. Most of these methods require plenty of hand labor to install them, and unless considerable care and fine craftsmanship are used, they leave much to be desired in the way of

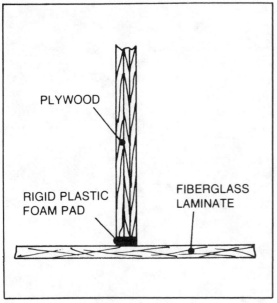

Fig. 9-23. Progressively wider bonding strips are used to taper the bond into the laminate.

appearance. Many require extensive sanding, a job that manufacturers especially try to avoid.

In an attempt to get around this problem, many manufacturers use *molded fiberglass liners*. These are used mainly with contact moldings that have only one molded finished side — the side that was in contact with the mold surface. The basic idea is to place the rough side of the main molding against the rough side of the molded fiberglass liner. Bond them together so that you have, in effect, a single laminate that is finished with a molded surface on both sides, similar to pressure moldings (see Chapter 8). Liners often make contact with and are bonded to main moldings only in certain areas. They perform other functions besides just giving a neat or finished appearance. They can also be used to stiffen, support, and reinforce the main molding; to form functional parts of the construction; and to provide a means of attachment for various things.

For example, a fiberglass auto body might have an interior head liner of molded fiberglass. This could give a neat and finished look; reinforce, stiffen, and strengthen the exterior molding; provide a means for mechanically attaching interior

118

lights without drilling holes through the external molding; and provide a place for running wires out of sight between the two moldings.

An interior liner for a fiberglass boat hull could give a neat and finished look; reinforce, stiffen, and strengthen the hull molding; provide a means for mechanically fastening plywood bulkheads and other structural members in place (reducing the amount of hand fiberglass lay-up work that must be done); and form the basic structure for the cabin sole or floor, berths, seats, galley, engine mounts, and so on.

The advantages of using molded liners, at least from a manufacturing point of view, are considerable. For example, think of the labor that would be required to construct all the things in the boat example by traditional methods. A single fiberglass molding eliminates most of this work.

A disadvantage from a manufacturing point of view is that you have to design and construct another expensive mold. Once the mold is constructed, production can usually be speeded up, sometimes greatly. The cost of the mold can be spread out or amortized by molding many units from the same mold.

Molded liners are becoming increasingly common in manufactured fiberglass boats. When fiberglass boats were first constructed, they were essentially fiberglass hulls with wood structural reinforcement, wood deck and cabin structures, and wood interiors — a wooden boat with the wood hull replaced with one of fiberglass. As time went on, wood decks and cabin structures were replaced with molded fiberglass ones. You then essentially had a molded fiberglass shell with a wood interior. Today, molded fiberglass liners are replacing much of the interior wood construction, with only small amounts of wood being used for trim, decoration, doors, rails, and other purposes.

This does not necessarily mean a better boat. A fiberglass liner inside a boat can give the appearance of the inside of an icebox unless large areas are covered with upholstery, wood, and other materials. If this is done, some of the advantages of using the hull liner, at least from a manufacturing point of view, are lost.

The most expensive custom boats generally have much less molded fiberglass showing on the interiors. Some boats are almost completely lined with wood. The boats that are produced in large numbers are the ones going to the liners.

Molded liners do require more molding work. They require less labor and materials than construction without the use of liners. After the liner has been molded, it is frequently set into the main molding before the main molding has been removed from its mold. Wet layers of fiberglass reinforcing material are often first placed in contact areas. The liner is then dropped into place. The fiberglass reinforcement and resin form a bond when the resin cures. This method is commonly used for boats and other applications where drilling holes through the moldings for mechanical fasteners is undesirable or must be kept to a minimum. For some applications, mechanical fastening is used in addition to or instead of bonding. Open spaces between the main molding and the molded liner are sometimes filled with rigid plastic foam or other materials to provide insulation, buoyancy floatation, core reinforcement between the two moldings, and other things.

The use of molded liners can make repair of damage to one or both moldings more difficult. If only one molding has a hole in it that must be repaired, for example, you might not be able to get to the area between the moldings for making the repair. The repair is usually made from one side only. A cutout can be made in the opposite molding, which also must be patched back into place after the first molding is repaired.

The use of molded liners does speed up some types of fiberglass construction. They are usually only used with contact molding, which gives only one finished side to a molding. As compared to, say, an automobile assembly line, most fiberglass manufacturing, even with the use of molded liners, is behind the times. This does not apply to some types of pressure molding production, some of which are automated to a greater degree than most American auto makers.

SUPPORTING FIBERGLASS MOLDINGS

Take care when handling, moving, and storing fiberglass moldings, especially those just taken from a mold. These are generally not fully cured (while polyester resin can cure to a hard state in a short period of time, the actual cure can go on for a period of weeks). Distortions can result from having a heavy load placed on one part of the molding. This sometimes results in a permanent distortion in the laminate.

To prevent this, moldings, especially new ones, must be properly supported so that they retain their original designed shape. This is especially important if stiffening and reinforcing members have not yet been added to the moldings.

For example, a boat hull should be placed on a cradle or support stand that properly supports the weight, so the hull laminate is not distorted (Fig. 9-25). To do this, the support load must be spread over a large area of the hull molding. Fiberglass boats carried on trailers that do not properly support the boats frequently leave permanent distortions in the bottoms of the boat hulls.

REINFORCING FIBERGLASS MOLDINGS FOR SUPPORTING HEAVY LOADS

This is a related but somewhat different problem than stiffening fiberglass laminates. For example, fiberglass boat and swimming pool decks must be able to support the weight of people walking on them and other loads that will be placed on them. A boat hull fiberglass molding must support the weight of a heavy engine. In these examples the load is downward, but it can be in any direction or combination of directions. The load on a cleat at-

Fig. 9-25. Fiberglass boat hull on a cradle.

tached to a fiberglass boat deck can stress the deck in a number of directions, depending on the angle of pull of a rope attached. Methods of attaching fittings to fiberglass moldings are covered later in this chapter. It is important to realize that thin fiberglass laminates must be reinforced if they are to support heavy loads.

The main principle involved is to spread the load over a wide area of the fiberglass molding. This is usually done in conjunction with stiffening the molding, so less distortion will occur to the molding when heavy loads are supported or applied. Fiberglass decks are often reinforced by thickening the laminate, backing the laminate with plywood, using core construction, or by means of frames and/or stringers, or some combination of these things.

In the case of the engine mounting in the fiberglass boat hull, the molding can be reinforced in the area where the engine is to be mounted. This can be done by thickening or adding extra layers to the laminate, and constructing engine mounts or the bed in such a way that the weight will be spread over a wide area of the hull. The mounts or bed also often attach to bulkheads and other major structural members.

The torque of the propeller turning in the water tends to twist the engine and the boat hull in the opposite direction. The hull and engine mounts must be able to withstand this. They must also be able to withstand the vibrations from the running engine, which tend to delaminate fiberglass bonds and cause cracking in heavy stress areas of fiberglass laminates.

Engine mounts and beds and other heavy load-supporting members are often integrated into molded fiberglass liners in an attempt to reduce the amount of labor required for construction and installation. The main idea is to make molded liners and components serve as many purposes as possible.

ASSEMBLING MOLDINGS

Fiberglass products should ideally be molded in one piece. Sometimes this is possible. Shower stalls, sinks, bathtubs, hot tubs, swimming pools, and a number of other fiberglass products can be molded in one piece. Some products require more than one molding or that a single molding be cut into separate pieces because of its design and functions. A fiberglass car body is an example. Either you make separate moldings for various components, such as the hood, doors, trunk lid, and main body, or you make a single molding and then cut out the hood, doors, and trunk lid from the main body component. The latter method, while it has been done, presents considerable difficulties in adding moldings and assembling the components and is almost never used in fiberglass manufacturing. It is usually much easier and better to use separate moldings that are designed for easy assembly.

It is impractical to mold some products in one piece, such as a boat with a deck and cabin structure. While a mold can be constructed that would allow removal of the molding from the mold by making the mold in sections that clamp together and can be taken apart to get the molding out, it is very difficult to do fiberglass laminating inside an enclosed tunnel-shaped mold. It is probably even more difficult to get workers who will do lay-up work under these conditions. Such a boat molded in one piece would probably be structurally superior to one molded in sections and then assembled by bonding and/or mechanical fasteners. Some attempts have been made to do this, including automated pressure molding that not only accomplishes this but also gives a molding with finished surfaces on both the exterior and interior sides.

This is still in the experimental stages, and most manufactured fiberglass boats are assembled from separate moldings. For example, the hull is one molding, and the deck and cabin structure is another. These are then joined together by bonding and/or mechanical fastening after at least one and usually both moldings have been removed from the molds.

Sometimes hulls are molded in two pieces and then joined together by fiberglassing bonding strips across the butt join of the two moldings (see Figs. 8-7 through 8-9). This is usually done to make molding easier. More of the lay-up work can be done in a near level position. Lay-up work in close quarters, such as at the bottom of a deep keel

section, is avoided, although someone has to apply bonding strips later in the same areas.

Boat hulls molded in this manner tend to break apart in the area where the two sections are joined. Even thick bonding lay-ups tend to be weaker than a single molding laid up in one piece. The bonding strips themselves can form a strong laminate with equal or even greater strength than that of the hull molding laminate. The problem is usually in bonding. The bonding laminate tends to peel away from or delaminate from the sections of the hull molding.

The strength of the bond can be greatly improved if bonding strips are added both inside and outside the hull instead of inside only. This requires difficult and time-consuming sanding and finishing work. Part of the molded finish is destroyed on the outside of the hull where the bonding strips are added.

The trend is toward molding hulls in one piece. To make this easier, the molds are often constructed so that they can be turned to any angle or rolled from side to side. The mold can then be positioned so that the lay-up work being done is on a level plane.

Several methods are used for joining moldings together: bonding by using strips of fiberglass reinforcing material and resin; mechanically fastening the moldings together by means of bolts, rivets, screws, and other fasteners; mechanically fastening the two moldings to a common joining member (the two moldings might or might not touch each other); gluing; and by some combination of these methods.

Bonding by Using Strips of Fiberglass Reinforcing Material and Resin

There are several ways of joining two moldings together. Most of these involve some form of a butt or lap joint of the two moldings. Figure 9-26 shows an end-to-end butt joint. An angle butt joint is shown in Fig. 9-27. Figure 9-28 shows a lap joint. By specially shaping the moldings in the molding process, these joints can often be improved. Figures 9-29 through 9-31 show some typical examples.

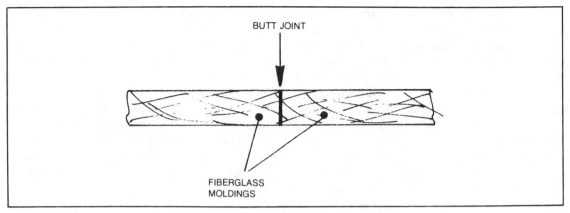

Fig. 9-26. End-to-end butt joint.

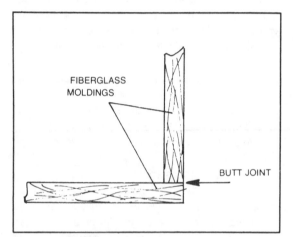

Fig. 9-27. An angle butt joint.

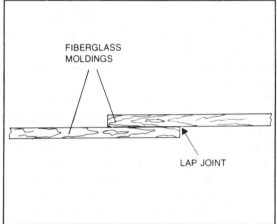

Fig. 9-28. A lap joint.

The moldings are positioned together. Bonding straps, angles, and fillets are laminated in place on one or both sides. Figure 9-32 shows an end-to-end butt joint with a bonding strap laminated in place on one side. Notice that progressively wider strips of reinforcing material are used. Figure 9-33 shows the same type of butt joint with a strap bonding laminate on both sides. If a smooth finish is required on one side, the moldings should be tapered on one side and filled in level with the bonding strap (Fig. 9-34).

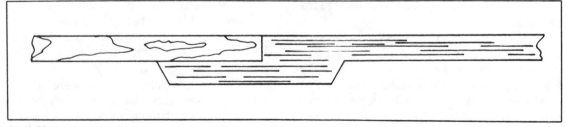

Fig. 9-29. Improved lap joint.

122

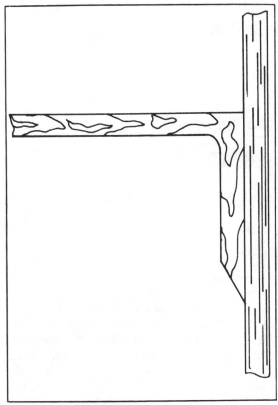

Fig. 9-30. Angle lap joint.

Figure 9-35 shows an angle butt joint with angle bonding straps laminated in place on one side. Notice that progressively wider strips of reinforcing material are used. Figure 9-36 shows the same type of angle butt joint with bonding angles laminated in place on both sides. If a smooth finish is required on one side, the moldings should be tapered on one side and filled in level with the bonding angle.

Fig. 9-31. Corner lap joint.

Figure 9-37 shows still another type of butt joint with the two laminates meeting in a T. Angle bonding straps are laminated in place, usually on both sides of the molding that makes edge contact.

Butt joints are sometimes used to make fiberglass tanks and other similar constructions from flat sheets of fiberglass. From a structural point of view, butt joints leave much to be desired, especially if moderate or heavy stresses will be placed on the joints. Also, these joints present difficult sanding and finishing jobs when appearance is important. It is the old problem of laminating over something instead of forming the desired finished side against a mold.

Butt joints are not often used for assembling manufactured fiberglass products when they must

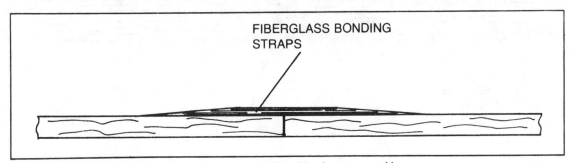

FIBERGLASS BONDING
STRAPS

Fig. 9-32. End-to-end butt joint with bonding straps laminated in place on one side.

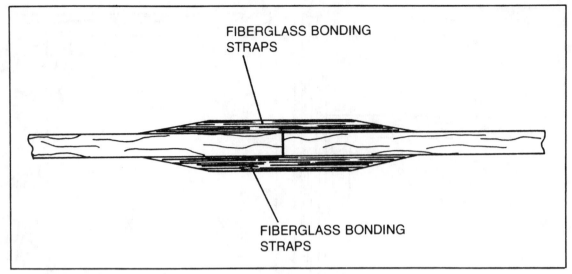

Fig. 9-33. End-to-end butt joint with bonding straps laminated in place on both sides.

be bonded together with bonding straps laminated in place. There are methods using aluminum extrusions that do not require fiberglass bonding that are used in manufacturing. These are detailed later in this chapter.

Lap joints are generally used instead of butt joints. These provide a larger contact area between the laminates. Wet resin with or without reinforcing material is often placed in the lap joint between the two moldings, which are then clamped or otherwise held in position until the resin cures. When used in this fashion, epoxy resin generally gives a much better bond than polyester resin.

The lap joints require additional reinforcing in

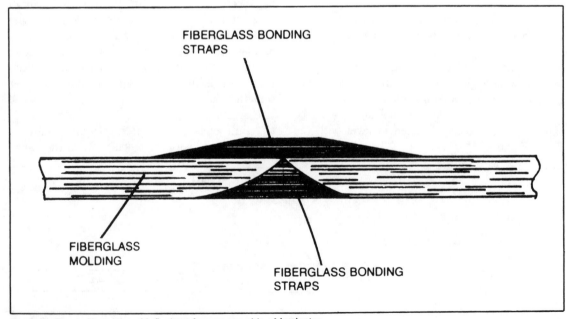

Fig. 9-34. Bonded butt joint with flush surface on one side of laminates.

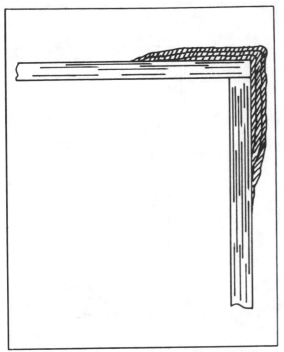

Fig. 9-35. Angle butt joint with fiberglass bonding straps on one side.

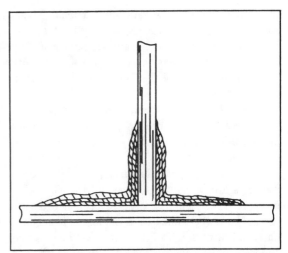

Fig. 9-37. Angle bonding straps joining two fiberglass laminates meeting in a T joint.

most cases. Lap joints can be mechanically fastened with through-bolts, as detailed later in this chapter.

Lap joints can also be reinforced with fiberglass straps and angles that are laminated in place. These can be on one or both sides of the laminates. When appearance is important, they are usually used on one side only. Sometimes a molding or rail made from wood, metal, or other material is used to cover up the joint on the finished side of the moldings (Fig. 9-38).

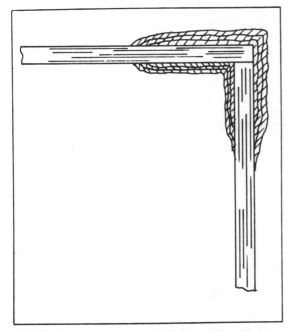

Fig. 9-36. Angle butt joint with fiberglass bonding straps on both sides.

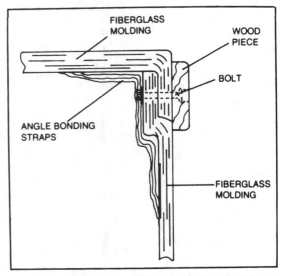

FIBERGLASS MOLDING

WOOD PIECE

BOLT

ANGLE BONDING STRAPS

FIBERGLASS MOLDING

Fig. 9-38. Angle lap joint with wood piece covering joint on the finished side of the molding.

Whenever possible, lap joints should be arranged so that the primary loads placed on the joints put them in compression, as shown in Fig. 9-39, rather than tension, as shown in Fig 9-40. Tension tends to pull the joint apart rather than push it together.

Butt and lap joints with sandwich core moldings present special problems. Figure 9-41 shows an end-to-end butt joint for joining two sandwich core moldings.

In fiberglass manufacturing, sandwich core moldings are frequently tapered into a single skin in areas where laminates are to be joined during the lay-up of the laminates in the mold. This eliminates the need for fiberglass or wood inserts and speeds up the construction. It also conceals and protects the core material. Lap joints such as those described earlier in this chapter for single-skin laminates can then be used for joining moldings.

It is also possible to join two sandwich core moldings or a sandwich core molding and a single-

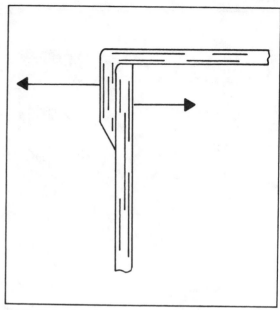

Fig. 9-40. Lap joint in tension, which tends to pull the joint apart.

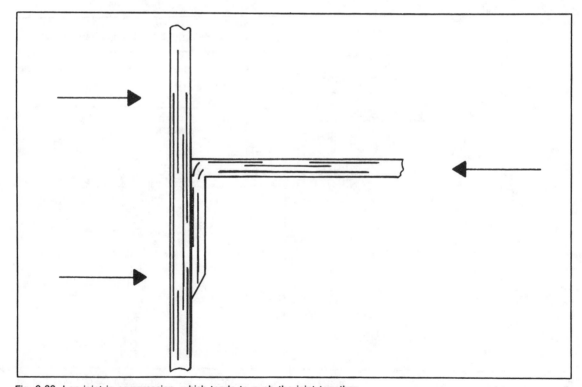

Fig. 9-39. Lap joint in compression, which tends to push the joint together.

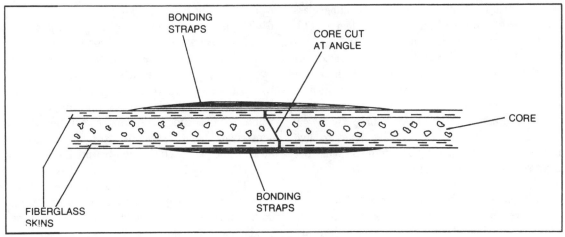

Fig. 9-41. End-to-end butt joint for joining two sandwich moldings together.

skin molding with angle butt joints. This generally isn't recommended except for light load applications.

Attaching Moldings Together by Means of Mechanical Fasteners

Bolts, rivets, and other types of mechanical fasteners can be used for attaching moldings together. These usually pass through holes drilled through the two moldings at lap joints, which might or might not also be bonded together. Special flanges for forming lap joints for mechanically fastening separate moldings together are often incorporated into the mold. Figure 9-42 shows a typical molded lap joint arrangement.

For maximum strength, the mechanical fasteners must be properly spaced. The correct spacing depends on the thickness of the laminates, required strength of the joint, and type of fasteners used. The fasteners must also be spaced a certain minimum distance from the edge of the moldings. Through-bolts and rivets are most frequently used. Washers or metal plates are often used with the fasteners to spread the load over a larger section of the moldings. Screws are sometimes used for joining moldings, but these are generally much less satisfactory than using through-bolts or rivets.

Sometimes these joints are arranged so that they can be taken apart by removing the fasteners and slipping the moldings apart. In other cases the joints are permanently bonded. The mechanical fasteners act as clamps until the bonding resin or glue sets up and reinforces the bond.

Sometimes the mechanical fasteners are used in conjunction with a wood or metal molding or rail (Fig. 9-38). A boat hull-to-deck attachment of fiberglass moldings is often accomplished with a lap joint, bonding between the contact area of the two laminates, through-fasteners that pass through the

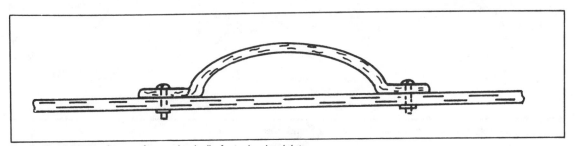

Fig. 9-42. Molded-in flanges for mechanically fastening lap joints.

127

wood or metal rail and the moldings, and fiberglass bonding straps on the inside of the moldings. This method of attaching hull and deck moldings has proved to be highly satisfactory. The wood or metal rail covers the edge of the moldings on the outside of the boat. No outside sanding or finishing is usually required. This method does require considerable production time to do it properly. Many designers and manufacturers have tried to find faster methods for making hull-to-deck joins, but most of these are much less satisfactory than the method described earlier.

Mechanical fasteners, especially bolts, can exert tremendous compression loads on small areas of the fiberglass moldings when they are tightened down. For this reason, take care not to overtighten them. Use washers and other backing pieces when possible to spread the load over a larger area of the laminates.

Mechanically Fastening Two Moldings to a Common Joining Member

A strip of wood can be positioned in the corner of an angle butt joint between two moldings. The two moldings can be attached to the wood strip with screws or bolts (Fig. 9-43). This type of joint is not very strong and is suitable only for joints that will have light loadings. A similar joint, but usually somewhat stronger, can be formed by using a metal angle piece and through-bolting the moldings to it (Fig. 9-44).

The two methods are seldom used in fiberglass manufacturing. They can be used by the do-it-yourselfer for assembling tanks and other similar items from flat sheets or panels of fiberglass.

An H-shaped aluminum extrusion is used by some fiberglass boat manufacturers for joining hull and deck moldings (Fig. 9-45). Notice that the

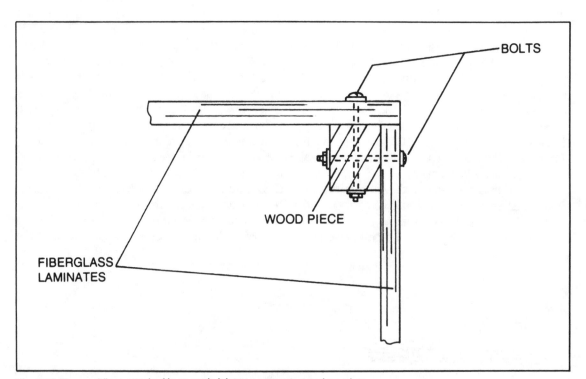

Fig. 9-43. Two moldings attached in an angle joint to a common wood member.

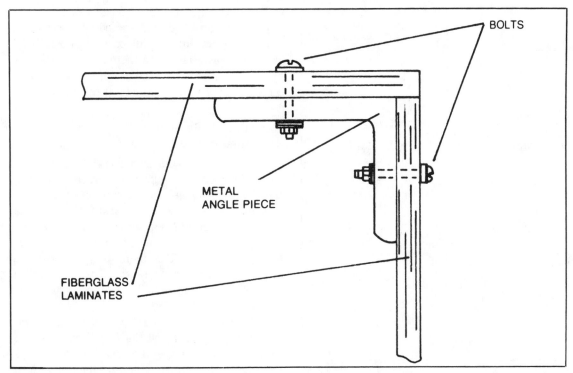

BOLTS

METAL
ANGLE PIECE

FIBERGLASS
LAMINATES

Fig. 9-44. Two moldings attached in an angle joint to a common metal angle piece.

edges of the two moldings do not make direct contact; they are separated by a section of aluminum. The two moldings are set in the tracks with bedding compound or sealer and then mechanically fastened in place, usually with rivets.

Several manufacturers made the claim that this method was as strong as, or even stronger than, the bonding and mechanically fastening method described earlier. Many failures of this joint on boats in use indicate otherwise. This method gained such a bad reputation that it is now difficult to market boats with this type of hull-to-deck connection.

The method certainly cut costs for manufacturers, at least before the lawsuits. It was cheap and fast. Unfortunately, the joint had the habit of coming apart, making the boats unsafe. Most of the manufacturers who tried this method have gone back to the more proven method previously described, or some variation of it.

Joining Moldings by Gluing

This is closely related to bonding by using epoxy resin (which is basically epoxy glue) without reinforcing material between contact areas of lap joints. While this alone is sufficient for some light-duty applications, the joints are usually reinforced by bonding strips of fiberglass laminated in place and/or mechanical fastenings.

Moldings are joined by gluing in some manufacturing applications, such as for attaching a honeycomb core of fiberglass or other material between two moldings. Under carefully controlled conditions, strong glue joints are possible.

MAKING ATTACHMENTS
TO FIBERGLASS MOLDINGS

It is necessary to attach various items to fiberglass moldings for many applications. For example, a shower stall requires a drain, water spout, and

129

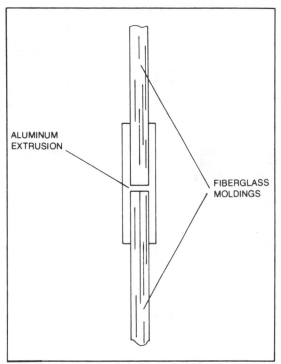

Fig. 9-45. An H-shaped aluminum extrusion used for joining fiberglass moldings.

faucets. A fiberglass auto body requires hinges and latches for the doors, hood, and trunk lids. A fiberglass boat requires attaching cleats and other fittings to fiberglass moldings. Several methods are used for attaching things to fiberglass moldings — the most common being bonding with fiberglass, mechanically fastening, and gluing.

Bonding with Fiberglass

Several applications of this method have already been described. For example, stiffeners are laminated in place, and bulkheads are attached with fiberglass angles laminated in place. This is a primary method of making attachments and is comparable to welding in metal construction.

This method allows strong attachments without making holes in the moldings for bolts and other fasteners. For example, a block of wood can be attached to a fiberglass molding by laminating it in place with fiberglass bonding straps or angles or even completely covering or embedding the wood

with fiberglass. After the laminate has cured, fittings can be attached to the wood with screws or other fasteners. Interior components in boat hulls are commonly attached to the fiberglass moldings by this method without the use of mechanical fasteners passing into or through the moldings.

Mechanically Fastening Attachments and Fittings

Mechanical fastening to fiberglass moldings is a commonly used technique in fiberglass construction. There are many situations where it is desirable to pass a metal or plastic fitting through a fiberglass molding. A shower pan or bathtub drain fitting is an example. A hole is drilled through the fiberglass molding. The drain fitting is positioned and set in caulking or sealer. The nut is then threaded on and tightened to hold the drain fitting in place.

Figure 9-46 shows a typical through-hull fitting on a boat. The installation is similar to that of a typical shower drain, except that a backing block (usually hardwood) is often added on the boat installation.

Metal hinges, latches, and other fittings are commonly attached to fiberglass auto bodies by bolts. Backing washers or plates are used to spread the load on the fasteners over a larger area of the fiberglass moldings.

Determine the directions and amounts of loads to be placed on the fasteners. The fiberglass moldings might require reinforcing. The loads placed on some fasteners will be in tension with the fasteners. The loads will be in compression in other cases. Large backing washers and plates are most important for tension loadings.

Figure 9-47 shows the attachment of a cleat to a fiberglass boat deck with the area of the laminate on the side opposite the cleat reinforced with a fiberglass laminate. Notice that this tapers gradually down to the thickness of the deck molding. A metal backing plate is used to spread the load over a larger area. Large backing washers might be sufficient, but the metal backing plate provides an extra safety margin.

Figure 9-48 shows a similar attachment using

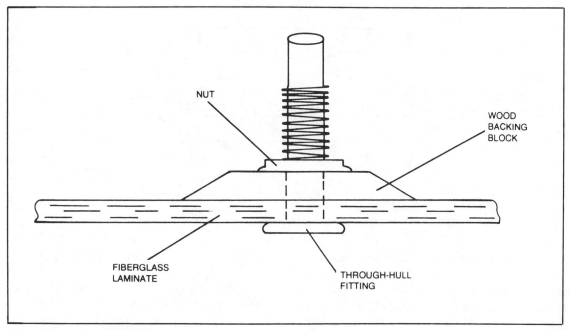

Fig. 9-46. Boat through-hull fitting.

a hardwood backing block instead of a fiberglass reinforcement. Large backing washers should be used.

Another possibility is to use a metal backing plate. Figure 9-49 shows a metal plate backing a chain plate on a sailboat. The plate is sheathed with fiberglass.

Sandwich core moldings with soft cores (end-grain balsa, rigid plastic foam, and so on) present special problems. A compression load on the fas-

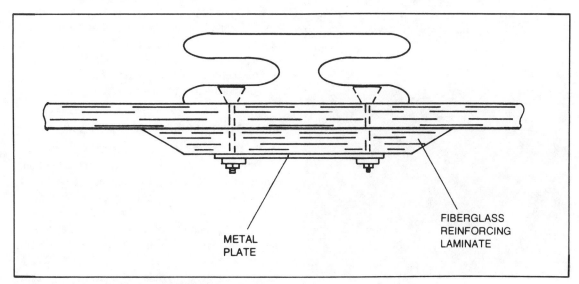

Fig. 9-47. Fiberglass molding reinforced with fiberglass laminate for attachment of a cleat.

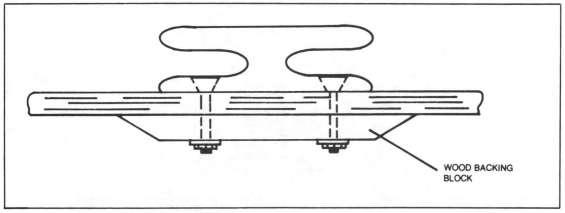

Fig. 9-48. Fiberglass molding reinforced with wood backing block for attaching a cleat.

tener forces the moldings inward, compressing the core material. To prevent this, several methods are in use. One method is to use inserts made from plastic, metal, or other materials (Fig. 9-50). In this way the insert rather than the core material takes the compression loading. Another method is to remove the core material in the area where the fastener is to be placed. Replace the core with fiberglass, which can be laminated in place (Fig. 9-51). Still another method is to remove a section of core material and replace it with plywood or some other material that is strong enough to withstand the compression loadings of the fasteners. To do this, a cutout can be made in one of the fiberglass skins. A section of core is then removed through the cutout. The new core is added, and a

fiberglass patch is laminated in place where the cutout was made.

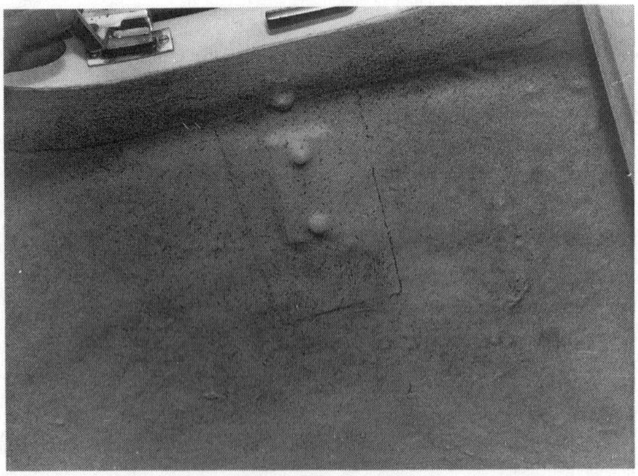

In some sandwich core manufacturing, it is known in advance where the fasteners will be positioned. Hard cores can be placed in these areas during the molding lay-up.

Attaching Things to Fiberglass Moldings by Gluing

This is another method used for attaching things to fiberglass moldings, especially for attachments that will have only light loadings placed on them. For example, a curtain rod bracket can be attached to a fiberglass molding by first epoxy gluing a block of wood to the fiberglass. The bracket can then be attached to the block of wood with screws. Rubber gaskets and weather strips are frequently attached to fiberglass by gluing.

PROTECTING FIBERGLASS MOLDINGS FROM WEAR AND ABRASION

Depending on the particular construction, certain areas of fiberglass moldings are usually more vulnerable to wear, abrasion, or damage than other areas. These areas can be wholly or partially protected in many cases. For example, metal bumpers are frequently used to protect the front and rear sections of fiberglass car bodies. Wood, metal, or plastic rub rails on boats are another example. Whenever anything rubs against a fiber-

Fig. 9-49. Metal backing plated covered with fiberglass.

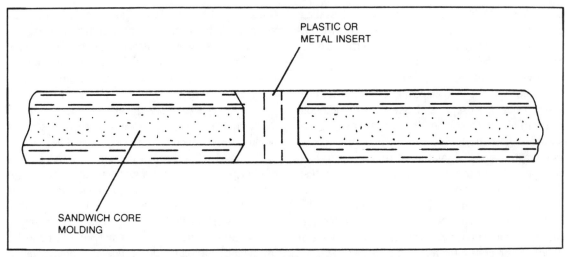

Fig. 9-50. Plastic or metal insert for a fastener to prevent crushing of the core material.

glass molding, wear or chafing is likely. Protect these areas with plates, strips, and moldings made from materials that have better abrasion resistance. The same applies to possible damage from impact. Protect the vulnerable areas with bumpers, shields, guards, and other means.

NONSLIPPERY SURFACES ON FIBERGLASS MOLDINGS

Smooth gel coat surfaces on fiberglass moldings tend to be extremely slippery, especially when they are wet. A nonslippery surface is desirable on

some fiberglass molding surfaces, such as swimming pool decks, shower stall floors, boat decks, and many other walking surfaces.

One method is to mold in a nonskid surface, such as a mat or waffle pattern, to the fiberglass laminate. This is a popular manufacturing method because nothing is added later. This method is frequently used for making nonskid surfaces on swimming pool decks, shower stall floors, boat decks, and other walking surfaces.

It is also possible to laminate on a nonskid matting or other material during the molding pro-

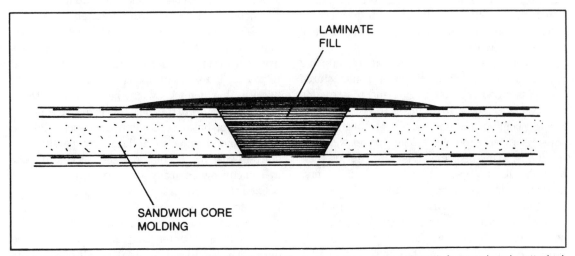

Fig. 9-51. Replacement of core material with fiberglass laminate in an area where a through fastener is to be attached.

cess. This method is used for some fiberglass manufacturing applications. A main difficulty with this method is in finding a nonskid material that is durable enough. Many materials will wear smooth quickly. Replacement can be difficult.

These methods are accomplished in the mold and molding process. It is also possible to add nonskid surfaces to existing moldings that have smooth surfaces. One method is to glue and/or mechanically fasten nonskid materials to the fiberglass, such as rubber, plastic, or metal treads, or teak decks that are sometimes used on fiberglass boats.

Still another method is to paint on a nonskid surface. Special paint-on compounds are manufactured, or you can add nonskid material to paint or polyester or epoxy resin.

MANUFACTURED FIBERGLASS PRODUCTS

For some fiberglass products, the thing that comes out of a mold is essentially a finished product. Sometimes additional trimming and/or minor assembly is required. For some products, the molding operation is only part of the manufacturing process; considerable stiffening, reinforcing, assembly, and attachments are required to produce a finished and marketable product.

Some products are all or nearly all fiberglass. Others are composites of materials such as rubber tires, metal chassis, and a fiberglass auto body with glass windows combined together to form a functional automobile. For these and other similar products, the fiberglass moldings are only part of the total assembly.

To make effective fiberglassing repairs, it's important to learn as much about manufactured fiberglass products, their design, and construction as possible. One way to do this is to visit manufacturing plants. Some offer tours, and many plant supervisors will show you around and answer your questions. Many people who go on to set up their own fiberglassing repair businesses begin by first working in fiberglass manufacturing.

Another way to learn about manufactured fiberglass products is to study them. When you see something made of fiberglass, observe not only how nice it looks but how it is constructed. How

thick are the moldings? What molding methods were used? How was stiffening and reinforcing accomplished? How were attachments made? Is there any sign of product failure? If so, does this appear to be a result of poor construction, design, or materials? Was it a result of misuse, abuse, or accident?

Also, try to compare similar products made by different manufacturers. Compare such factors as quality, appearance, and price.

Important factors in determining the quality of fiberglass products include: soundness of design, quality of materials used in the construction, and care and skill in fabrication. Remember that most fiberglass manufacturing companies are in business to make a profit, and this is not always done by producing high-quality products. Sometimes a lower quality product that can be marketed for a lower cost means more profit to the manufacturers. Fiberglass products are not created equally.

Soundness of Design

To produce a quality product, a first requirement is a sound design. For almost every kind and type of manufactured fiberglass product, there are examples of good, poor, and indifferent designs. As manufacturing businesses go, fiberglass production can be a fairly inexpensive one to get started in if you use hand lay-up in contact molds. You can construct or purchase the molds fairly inexpensively. Pressure molding and automated fiberglass production requires a huge investment.

Many people who jump into the fiberglass manufacturing business on a low investment basis use poor designs, either because they can't afford to pay for the services of a competent designer or because they don't know any better. Unless these people switch to better designs, going broke or at least out of business is often the result. On the other hand, I know of at least one person who started what has become a highly successful boat manufacturing company on very little money. I believe one of the reasons for success is that he had an outstanding and proven design.

In some cases the name of the designer will be important, as is often the case with boats. The designer's name will be used in advertising and promoting the boat. In other cases the name of the designer will not be important. Who knows the names of any fiberglass bathtub designers?

Design is often thought of only as the way something will look — the shape of the thing. Design also includes function or how it will work (how well a boat will sail, for example) and how it will be constructed (materials to be used, number of layers in laminates, and so on). Some things, especially construction, can be left up to the manufacturer. Most designers are proud of their creations. They do not like manufacturers to make changes without their approval or to lower the quality by skimping on materials.

Quality of Materials Used in the Construction

Many of the materials used in fiberglass construction are more or less hidden from view in the finished products. Thus, the temptation to use lower quality materials to lower production costs is always present. This is especially true today with the high cost of resin and other materials used in fiberglass construction. While most fiberglass products are engineered so that they are as strong as necessary, plus a safety margin, there is a trend toward reducing the safety margin — sometimes to the point where the products won't hold up or are even dangerous.

Some manufacturers have built up their reputations on producing quality products. While the costs of most fiberglassing materials are presently at a record high, improved formulas of resins, better reinforcing materials, and other material improvements sometimes make it possible to use less materials without lowering quality.

The selling price of similar manufactured products often reflects the quality of the materials used in the construction — that is, you get what you pay for. The same product by the same manufacturer often sells for various prices, depending on where you purchase it, how much discount you can get, and so on.

The quality of the materials used in the original construction will have a bearing on repairing damage to it later. As a rule, quality products will suffer less damage (for example, from equal impact) and be easier to repair when they are damaged.

Care and Skill in Fabrication

This is another important factor in determining the quality of manufactured products. Both materials and labor are presently expensive. While care in fabrication has always been a problem in fiberglass manufacturing from the time the first products were manufactured, it's especially important today.

It is extremely difficult to get skilled fiberglass workers, especially for some of the less desirable jobs such as lay-up and spraying work. These jobs tend to be unpleasant and present health risks to the workers. Improved industrial health and safety standards and regulations have improved the working conditions somewhat and reduced the health risks, but manufacturing costs have also increased.

Manufacturers have used many methods to reduce the amount of labor required for fiberglass production. Some of these methods, such as using spray-up with a chopper gun instead of hand lay-up and the use of molded fiberglass liners, have already been mentioned. While these methods do not necessarily increase quality, and in some cases might even lower it, they can reduce the amount of labor required.

Automated fiberglass production is probably the wave of the future. It has been mainly used so far for fairly small moldings, and even these are seldom fully automated. The trend is definitely toward automation and to apply this to larger moldings and products.

Chapter 10

Repairing Fiberglass

Methods in this chapter are given for repairing various types and degrees of damage to fiberglass. It is assumed that you are thoroughly familiar with the material covered in the book to this point, that you will observe safety rules and precautions for working with fiberglassing materials and chemicals, and that you have done all the practice exercises in the polyester resin section and also the epoxy resin section of Chapter 7, if epoxy resin is to be used for repair work.

WORKING CONDITIONS

Fiberglassing repair work should be done, ideally, under laboratory conditions where factors such as temperature and humidity can be controlled. Obviously, most fiberglass repair work will have to be done with less than ideal conditions. Try to do the most crucial work, such as laying up fiberglass laminates, with the best possible working conditions. About 75 degrees Fahrenheit seems to be best for applying most polyester resins, but by adding more catalyst, you can do satisfactory work with regular polyester resins from about 60 degrees Fahrenheit. By adding less catalyst, you can easily work with polyester resins up to about 95 degrees Fahrenheit. Special low-temperature polyester resins are on the market that allow working in the 45- to 60-degree range, but these resins are usually quite difficult to work with. Another possibility for working in cold temperatures is to use a heat lamp or other flameless heating device.

Try to avoid damp and humid days. Working under these conditions can result in poor bonding and improper curing of resins.

If possible, avoid fiberglassing in direct sunlight. Some type of sunshade can be rigged up in many cases. Although it's especially uncomfortable to wear safety clothing and equipment in hot weather, either wear it or wait for cooler working temperatures.

In the practice exercises given in Chapter 7, the fiberglassing was done on a level plane. In many cases this cannot be arranged for repair work, and fiberglassing will have to be done on vertical and even overhead surfaces. Special high-viscosity polyester resins with a thixotropic agent added are available. These resins drip, run, and sag less than regular viscosity polyester resin and make it possible to do satisfactory fiberglassing on vertical and overhead surfaces.

Fiberglassing repair work must be done under the cleanest possible conditions. Dust, dirt, and grime can mix in with fiberglassing materials and chemicals and reduce the quality of the resulting laminates. When doing repair work on location, clean up the working area as much as possible before starting fiberglassing.

MINOR REPAIRS

Fiberglass can be damaged in many ways, such as by wear, abrasion, impact, sun exposure, and so on. The damage can range from minor to major. It can be cosmetic, structural, or a combination of these.

An important fiberglassing repair skill is making minor repairs to fiberglass moldings. While these are considered minor, it still requires considerable practice and skill to do them well, especially when application and matching of gel coat and paint finishes are involved.

Scratches and Minor Gouges

First consider scratches and minor gouges in gel coat surfaces that cannot be removed by polishing and buffing, as detailed in Chapter 5.

If the scratch or gouge does not go through the gel coat, repair as follows. Using a clean white cloth saturated with acetone, wipe the area of the scratch or gouge. Colored cloth should not be used, as the acetone will often act as a solvent for the dye and stain the gel coat of the molding being repaired. Use light pressure only. Rubbing too hard with acetone will remove some of the gel coat. The purpose of the cleaning is to remove dirt and other loose particles from the surface, so they will not cause additional scratching during the rest of the repair.

Using a sanding block and 220-grit abrasive paper, carefully sand until the scratch or gouge disappears. Do not remove any more of the gel coat than is absolutely necessary.

Thoroughly clean the area with a cloth soaked with water. Wet sand with a sanding block and 400-grit sandpaper. Then wet sand with a sanding block and 600-grit sandpaper.

Polish the surface and remove any remaining scratches by using rubbing compound. This can be applied by hand with a cloth, or a portable electric polisher with a buffing pad can be used.

If in the process of the repair you go completely through the gel coat, touch up the gel coat surface as detailed later in this chapter. This will sometimes happen even when you are careful, as gel coats are usually quite thin.

If the scratch or gouge is on a painted fiberglass surface, the method of repair will depend on the type of paint used. Sometimes epoxy and polyurethane paints can be repaired in the same way as gel coats. Scratches and gouges can sometimes be removed from other types of paint with polishing and rubbing compounds. Try these first in an area where it does not show. Sanding and touch-up painting will be required in some cases.

Scratches and minor gouges that go below the gel coat or paint usually require filling with fiberglass putty. While either polyester or epoxy putty compounds can be used, epoxy generally gives better results. To make a repair, proceed as follows.

Clean out scratches before applying the putty is applied. Use a sharp-pointed metal object and/or a folded-over piece of 100-grit sandpaper. Similarly, scrape out and/or sand gouges.

On gel coat surfaces, clean out the scratch or gouge with acetone on a clean white cloth. On painted surfaces, wipe the area clean with a cloth. Do not use acetone unless you first test the paint with acetone in an area that does not show to make certain that acetone does not act as a solvent to the paint.

The choice of putty depends on many factors. Special putty compounds are available with gel coat colors already mixed in or in kits with necessary pigments for color matching. This will usually eliminate the need for touching up the area with gel coat later. If the original gel coat is white, manufactured polyester and epoxy putty compounds that are white will often give adequate color matching. In some cases a thixotropic thickening agent and a color pigment are added to polyester or epoxy resin to form a filling compound. These will sometimes work satisfactorily for filling very small areas, although the compounds tend to form a somewhat brittle putty that might tend to crack or even fall out.

If manufactured compounds are used, follow the mixing and application instructions on the labels carefully. Measure out the required amount of putty and add catalyst or hardener. Mix thoroughly.

Use a putty knife to apply putty. Fill the scratch or gouge. Use a putty knife as a drag to remove excess putty. If polyester putty is used, slightly overfill to allow for shrinkage.

After the putty has cured, sand off excess putty and smooth out the surface. Use a sanding block. Start with 100-grit sandpaper. Follow this with 220-grit abrasive paper.

If color has been mixed in with filler, finish by wet sanding with 400-grit and 600-grit sandpaper. Then polish the surface and remove any remaining scratches by using rubbing compound. This can be applied by hand with a clean cloth or use a portable electric polisher with a buffing pad.

If the fiberglass molding being repaired has a gel coat surface, and your repair requires gel coat touch-up work, you will need gel coat resin in matching color. Gel coat kits are available with necessary pigments for color matching. Follow the manufacturer's directions for mixing and application. Brush on or spray the gel coat. Apply a thin layer, usually from about 10 to 15 mils in thickness. Avoid too thick a layer as it will then tend to crack after it has cured.

If the gel coat resin is air-inhibited, seal it off from the air to get the gel coat surface to cure tack-free. One way to do this is to cover the area where the gel coat has been applied with cellophane. Tape the cellophane in place around the edges with masking tape beyond the area where the gel coat has been applied. Another method is to brush or spray polyvinyl alcohol (PVA) over the resin. This will seal the resin off from the air and allow it to cure.

Imperfections in the gel coat touch-up can be corrected after the resin has fully cured by wet sanding first with 400-grit and then 600-grit sandpaper. Use rubbing compound for surface scratches.

On painted surfaces, touch up the repaired area with a matching type and color of paint. A primer might be required, depending on the particular paint.

Nicks and Small Holes

Nicks and small holes that do not pass all the way through the fiberglass molding can be repaired in a similar manner. Prepare the surface by sanding. Use coarse enough sandpaper to remove any loose material. An abrasive grinding burr attachment in a portable electric drill can also be used for preparing the surface of the nick or small hole.

On gel coat laminates, clean out nicks and small holes by wiping with acetone-soaked cloth. On painted surfaces, wipe with a clean cloth. Do not use acetone without first testing it in an inconspicuous area for colorfastness.

Small defects up to ¼ inch or so across and deep can usually be filled with polyester or epoxy putty, as detailed earlier for scratches or minor gouges that pass through the gel coat or paint surfaces. For filling larger areas, cut a piece of fiberglass mat to fit the damaged area. Depending on the depth of the damaged area, additional layers might be required. One layer of 1½-ounce-per-square-foot fiberglass mat will give a laminate about 1/20 inch thick. Catalyze an approximate amount of polyester resin that will be required to bond and wet out the reinforcing material. Epoxy resin can be used instead of polyester resin. Using a small brush, apply a layer of resin to the area to be repaired. Place the first layer of mat on the wet resin and press it in place. The side of the mat that joins the wet resin can first be saturated with resin. Using the brush, thoroughly saturate the mat with resin.

Additional layers of mat, if required, can usually be applied immediately. You can allow the first layer to cure before applying the second, and so on. Usually two or three layers of mat can be applied to a small area in one operation. If more layers are required, it's best to allow these to first cure before adding additional layers. It is difficult to keep a proper resin-to-reinforcing-material ratio if more layers are attempted at once. The repair laminate should be slightly higher than the surrounding molding. When everything looks okay, clean the brush with acetone for polyester resin and with epoxy solvent for epoxy before the resin has a chance to harden on the brush.

Allow the putty or resin to cure. If polyester laminating resin was used, apply a thin coat of finishing resin to the surface, so the laminate will cure with a tack-free surface. This must be done before attempting sanding.

Remove excess material by sanding. Use a sanding block. In most cases 100-grit sandpaper will be about right, but some jobs might require coarser grits for removing excess material in a reasonable amount of time. Make certain, however, that sanding scratches do not go below the desired finished surface. When the desired surface level is close, switch to 220-grit paper. Touch up the finish with gel coat or paint.

Small Cracks and Crazing

This section applies only to moldings that have a gel coat. Small cracks and crazing (hairline breaks of random pattern) in gel coat are often the result of improper application or formulation of gel coat, especially from applying too thick a layer of gel coat. This is especially a problem with older moldings. The formulation of gel coat resin has been improved in recent years. If this new resin was used, cracking and crazing is less likely.

Cracking and crazing of the gel coat is also caused by localized stresses on the moldings, such as from fasteners that are too tight or by working of the laminate at stress points. This section concerns cracking and crazing in the gel coat only. Deeper cracks and fractures that extend into or through the laminate below require additional repair and are covered later in this chapter.

To make a repair, proceed as follows. First, examine the cracking and crazing and try to determine the underlying cause. If possible, correct the problem. For example, if fasteners are too tight, loosen them. If this is not done, new cracking and crazing of the gel coat is likely after the repair has been made.

The cracks and crazing usually extend all the way through the gel coat. If not, repair can be made in the same way as for scratches and minor gouges that do not extend all the way through the gel coat.

To apply polyester or epoxy putty, widen the cracks in the gel coat. Use a sharp-pointed metal object. If this isn't done, it will be difficult to get putty down into the cracks.

Using a clean white cloth saturated with acetone, clean the area. Fill cracks with polyester or epoxy putty. Use a putty knife to apply the putty and as a drag for leveling the surface. If polyester putty is used, slightly overfill to allow for shrinkage during curing.

After the putty has cured, sand off excess. Use a sanding block and start with 100-grit or finer (higher grit number) sandpaper. Follow this with 220-grit sandpaper. Try not to scratch the surrounding gel coat.

If color has been mixed in with the putty, finish by wet sanding with first 400-grit and then 600-grit sandpaper. Polish the surface and remove any remaining scratches by using rubbing compound by hand with a clean cloth or with portable electric polisher with a buffing pad.

If matching color was not mixed in the putty, apply matching gel coat. Allow the gel coat to cure to a tack-free surface. Imperfections in the gel coat touch-up can be corrected by wet sanding first with 400-grit and then 600-grit sandpaper. Then use rubbing compound.

Filling in Small Holes Through Fiberglass

There are many cases where you need to fill in small holes that have been drilled through fiberglass. It is a lot easier to drill holes in fiberglass than it is to fill them back in. While it is usually a fairly minor fiberglass repair to fill in small holes (my concern here is with holes up to about ½ inch in diameter—repair of larger holes is covered later in this chapter), it would have been better to have avoided making them in the first place.

While small holes can be filled in merely by applying polyester or epoxy putty, a more substantial repair is usually needed. If you can get to both sides of the hole, "V" the hole out on both sides to the center (Fig. 10-1). Extend the taper approximately half the diameter of the hole or more away from the edge of the hole to give a good bonding area. Use a file or abrasive burr attachment in a portable electric drill.

If you can only get to one side of the hole, prepare the hole in a similar fashion. Use a small taper angle on the back side (Fig. 10-2). A small round (rat-tail) file can be used.

Apply a backing to one side of the hole. If you can get to both sides, use masking tape. Apply it so that it covers the taper on one side of the hole and extends across the center of the hole halfway through the molding. Apply it to the finished or gel

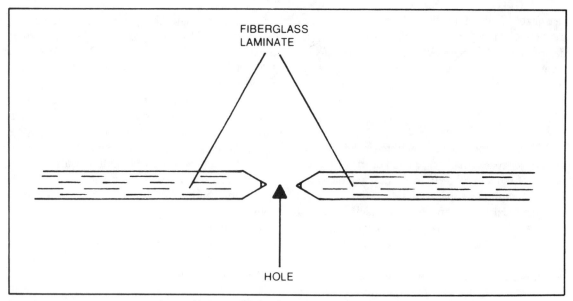

Fig. 10-1. Preparation of laminate for patching a small hole when you can get to both sides of the laminate.

coat side first, if you will be able to fiberglass from the back side of the molding. If you can only fiberglass from one side, apply the backing straight across the hole on the back side (Fig. 10-3). If the backing must be applied through the hole, stuff a thick mixture of polyester putty and chopped strands of fiberglass reinforcing material through small holes and pull it back into position to form a backing by using a bent piece of small diameter wire (Fig. 10-4). For holes from about ⅜ to ½ inch in diameter, cut a piece of cardboard to a diameter about twice that of the hole. Then attach a small diameter wire to the center of the cardboard. Roll

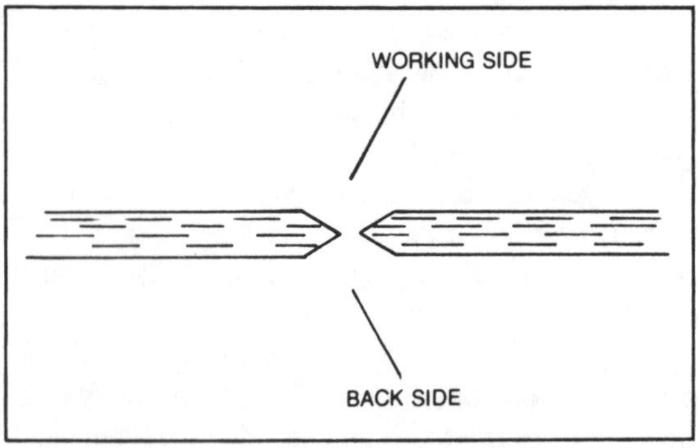

![Fig. 10-2 diagram with WORKING SIDE and BACK SIDE labels]

Fig. 10-2. Preparation of the hole for patching when you can only get to one side of the laminate.

the piece of cardboard up, force it through the hole, allow it to unroll, and pull it in backing position with the wire (Fig. 10-5).

Catalyze polyester resin and saturate chopped strands of fiberglass or small pieces of fiberglass mat with resin. Then fill half or all (depending on where the backing was placed) of the hole with this material. If you can fiberglass from the back side and a flush patch isn't necessary, add one or more layers of mat that extend beyond the hole being filled ½ inch or more. Saturate these with resin (Fig. 10-6). If you can fiberglass only from one side with flush backing, completely fill the hole. If half of the hole was filled, allow the resin to cure, remove the backing, and fill the other half of the hole.

Sand off excess material on the finished side of the molding. Use a sanding block. Usually only one side will be required to have a finished surface, as the back side of the laminate will not show. In a few cases a finished surface is required on both sides of the molding. Sand both sides. Gradually work down to finer grits (higher numbers) of sandpaper. Touch up the gel coat or paint as required.

This method will usually give a repair that is at least as strong as the original molding surrounding the patched hole provided that the bonding surface is properly prepared and cleaned, quality

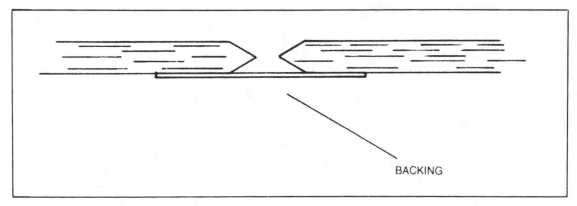

Fig. 10-3. Application of backing.

fiberglassing materials are used, and proper bonding and laminating techniques are used. An alternate method is simply to fill the holes with polyester or epoxy putty, but this often doesn't give a repair that is as strong as the original molding surrounding the patched hole. If this type of patch is used, it's good practice to use a mat laminate backing that extends ½ inch or more beyond the hole being filled to reinforce the putty bond.

Patching Larger Holes Through Fiberglass

Patching larger holes, from about ½ inch to several inches or more in diameter, is not much more difficult than patching small holes. I suggest that you forget about using putty to fill these and lay up a fiberglass laminate instead. This results in a much stronger repair.

If you can get to both sides of the hole and can fiberglass on both sides, taper laminate on both sides to the center of the laminate (Fig. 10-1). If the fiberglassing is to be done from one side of the hole only because you can only get to one side, or it would be too difficult to fiberglass there, taper the hole (Fig. 10-2). This job can be done with a file, an abrasive burr attachment in an electric drill or, for larger holes, a disk sander. The tapers give greater bonding area for the fiberglass laminate patch and also form a wedge to hold it in place. It's also possible to taper the hole on one side only, usually the finished side, especially if a backing laminate can be used (Fig. 10-7).

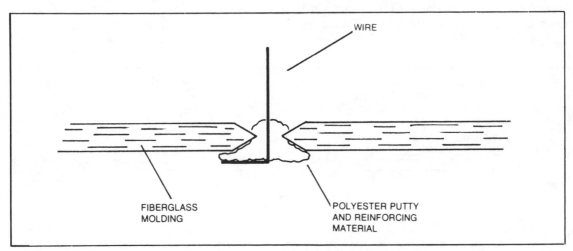

Fig. 10-4. Using wire with a bent end for applying putty backing through the hole.

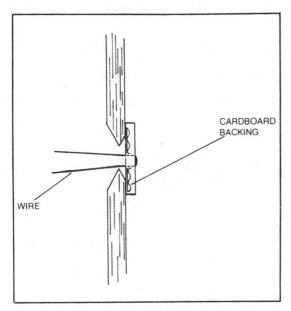

Fig. 10-5. Cardboard backing installed through the hole and pulled back in place with the wire.

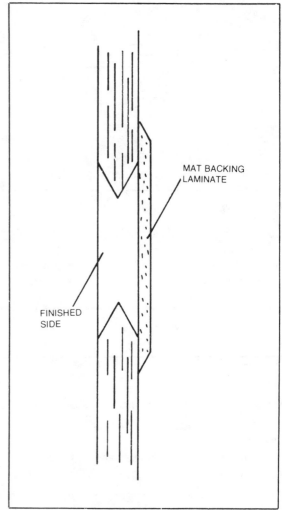

Fig. 10-6. Mat backing laminated in place.

Apply backing. This can be cardboard, wood, metal, or other materials. Hold it in place with tape, props, clamps, or by other means. If you are fiberglassing from both sides, apply the backing first over the finished side to the center of the taper (Fig. 10-8). If you are fiberglassing from one side only, place the backing flush with the back of the molding. If you cannot get to the backside of the hole to attach the backing, install the backing through the hole by cutting a piece of cardboard larger than the hole and attaching small diameter wire to the center of the cardboard. Roll the card-

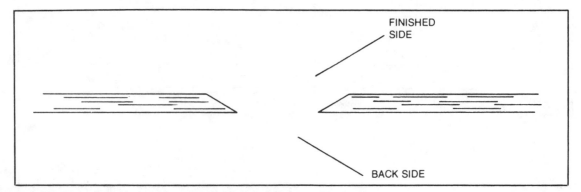

Fig. 10-7. Taper of hole for patching when backing can be laminated in place.

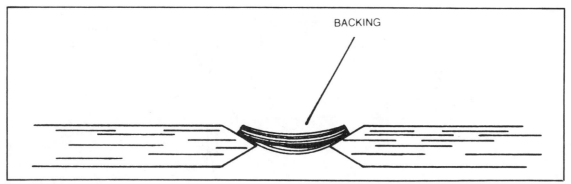

BACKING

Fig. 10-8. Application of backing for patching half of the hole.

board up and pass it through the hole. Pull it into position by wire. Then attach wire to a stick placed across the front of the hole to hold the backing in position (Fig. 10-9). It is also possible to install a mat backing by first placing mat saturated with high-viscosity polyester resin on the cardboard with the wire passing through the center of it. Before the resin has a chance to harden, roll the cardboard up along with the mat enough so that you can pass it through the hole. Pull the mat and cardboard into position behind the hole with the wire. Allow resin to cure, then cut off the wire that extends through the cardboard flush with the cardboard.

If you are fiberglassing from both sides, laminate the back half of the patch in place using layers of fiberglass mat (or, if desired, alternate layers of mat and cloth). Precut the reinforcing material to size. If the back side of the laminate does not show, add fiberglass backing that extends well beyond the edges of the hole. One or more layers of mat should be adequate for most patches, but cloth and/or woven roving can also be used if extra reinforcement or strength is required for a particular patch. Before adding this backing laminate, remove all paint from the bonding area. Clean the area with a clean white cloth saturated with acetone. Allow this to evaporate before fiberglassing. Allow the first half of the patch to cure, then remove the cardboard backing or other material. If care is taken not to get release agent on the bonding surface, apply the release agent to the backing before laminating in the first half of the repair for

easy removal of cardboard backing. If this is not done, laminate will also bond to cardboard or other backing, making removal of the backing more difficult. If release agent is not used, peel the cardboard off as much as possible. Scraped or sand off the remainder, then wash the area with acetone. Apply the second half of the patch, using precut pieces of mat and/or other reinforcing material.

If you are fiberglassing from one side, clean the bonding area with acetone. Laminate the patch in place using precut pieces of fiberglass mat and/or other reinforcing material. Generally, mat will be used if the original laminate is all mat. If the original laminate contained other reinforcing materials, you might want to try to duplicate the original lay-up, especially for larger size holes. The basic principle is to make the patch at least as strong as the original laminate and, if possible, stronger. If the original laminate is fairly thin, say, ⅛ inch thick or less, the entire laminate can usually be done in one operation. For thicker patches, lay up a ⅛-inch or less thickness and allow this to cure before adding another ⅛-inch thickness, and so on. After you gain experience, you will know approximately how thick a laminate you can lay up in one operation without getting into difficulty. When the patch extends slightly above surface level, allow the resin to cure.

Sand off excess material. If polyester laminating resin was used, it will first be necessary to apply a thin coat of finishing resin so that the surface will cure tack-free for sanding. You can use some other method for sealing the surface of the

143

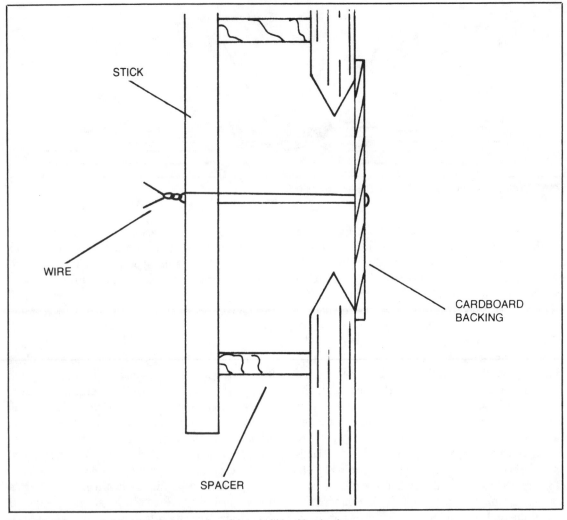

STICK

WIRE

CARDBOARD
BACKING

SPACER

Fig. 10-9. Wires are twisted together around a stick to hold backing in place.

resin from the air so that the surface will cure tack-free. Use a sanding block and begin with fairly coarse grit sandpaper. Usually only one side will show and require sanding and finishing. Sometimes a finished surface will be required on both sides. When most of the excess material has been removed, work down to finer (higher grit number) sandpaper. Touch up the gel coat or paint as required, matching the original finish as closely as possible.

Repairing Minor Damage
to Sandwich Core Moldings

Repair of minor damage that does not extend all the way through one of the fiberglass skins can usually be accomplished by using the same methods as for single-skin laminates. Minor damage that goes through one skin, or through one skin and the core, but not into the second skin, is usually repairable by the following steps.

Broken, cracked, and weakened fiberglass

144

should be sanded or ground back to solid laminate. Then taper the hole as shown in Fig. 10-10 by sanding or grinding.

Repair any damage to the core. Fill small areas of wood and plywood cores with polyester or epoxy putty. Rigid plastic foam and balsa cores can usually be filled in the same way. *Microballoons* (microscopic hollow balloons filled with nitrogen) can be added to polyester or epoxy resin to form a lightweight filler material. Still another method is to remove a section of the core and splice in a filler piece of the same material. This can be quite difficult and is not usually necessary.

The core will now serve as a backing for laminating a repair section to the damaged fiberglass skin. Precut the fiberglass mat and/or other reinforcing material. Brush on a layer of resin to the core and bonding areas of the original laminate. Apply the first layer of reinforcing material. This can be put in place dry, or it can be first saturated with resin. Saturate the reinforcing material with an even layer of resin. Add additional layers to the laminate. This can often be done in a single operation on high laminates. Apply in sections on thicker laminates, allowing each to cure before laminating the next one in place.

When the laminate is complete, allow it to cure. If polyester laminating resin was used, brush on a thin layer of finishing resin so that the surface will cure tack-free for sanding.

Sand off excess material. If you are hand sanding, use a sanding block. A disk power sander can also be used. Begin with fairly coarse grit sandpaper but not so coarse that scratches will extend below the desired finished surface. When most of the excess material has been removed, work down to finer (higher grit number) sandpaper.

Touch up the gel coat or paint as required. Match the original finish as closely as possible.

Minor damage that extends through both fiberglass skins is usually repairable by the following steps. Broken, cracked, and weakened fiberglass should be sanded or ground back to solid laminate on both skins. Then taper both skins as shown in

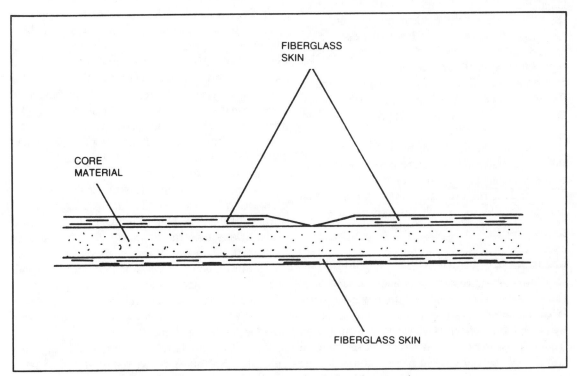

Fig. 10-10. Taper the edge around the damaged area in preparation for applying a patch.

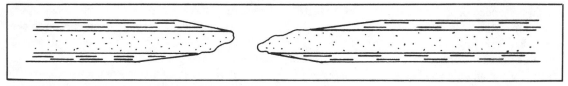

Fig. 10-11. Preparation of sandwich core laminate for repair.

Fig. 10-11 for bonding on patches by sanding or grinding.

Repair any damage to the core. Fill small areas of wood and plywood cores with polyester or epoxy putty. Fill rigid plastic foam and balsa cores in the same way add or microballoons to polyester or epoxy resin to form a lightweight filler material. You can also remove a section of the core and splice in a filler piece of the same material. This can be quite difficult and is usually not really necessary.

The core will then serve as a backing for laminating in the two patches. Laminate these in place one at a time.

If you can only fiberglass on one side of the laminate, cut out the damaged area through both laminates and the core. Then laminate in a patch on the back skin using the method given in this chapter for single-skin molding. Repair the core as described earlier. The core will then act as a backing for laminating in the second patch.

Allow laminates to cure. If polyester laminating resin was used, brush on a thin layer of finishing resin on surfaces to be sanded so that they will cure tack-free.

Sand off the excess material. When most of the excess material has been removed, work down to finer (higher grit number) sandpaper.

Touch up the gel coat or paint as required on one or both skins. Match the original finish as closely as possible.

MORE EXTENSIVE REPAIRS TO LAMINATES

More extensive damage to fiberglass moldings can be either easy or difficult to repair. A fairly large fracture or hole in a molding is not necessarily much more difficult to repair than a small one. There is just more of it. A larger repair laminate will have to be laid up, and more material will be required for doing it.

While almost any imaginable damage could probably be repaired, there is a definite limit as to what is practical to repair. In each specific case a decision should be made as to whether or not it is worth the cost and labor involved to make the repair and to decide if you have the necessary skill and experience to do the work. I've seen cases where a third or more of the moldings on a boat had to be replaced. While this was sometimes more or less successfully done, the cost and amount of time it took to do the job often seemed unwarranted.

Repairing Fractures and Holes in Single-Skin Laminates

Impact and a variety of other conditions can cause rupturing of fiberglass laminates. The flexibility of fiberglass is only about one-eighth that of steel, and fiberglass only stretches a limited amount before a rupture will occur. Repairs for this and related types of damage need to be functional and, in many cases, also aesthetically pleasing. Because it is difficult to be certain of the quality of a repair laminate, it should generally be thicker and stronger than the molding being repaired to give an adequate safety margin.

Steps for making a typical repair are as follows. Carefully survey the damaged area. Whenever possible, examine the molding from both sides. Determine the limits of the damaged area and mark off this area with chalk or marker pen. This can be done by visual inspection and sound, such as by tapping the area surrounding the damage lightly with a small metal hammer. A damaged area of a fiberglass laminate gives a different sound than an undamaged area.

In most cases broken, weak, and delaminated fiberglass will have to be removed before laminating in a repair. Mark the area to be cut out with chalk or marker pen. This should be on or outside the damaged area that was marked off earlier and should have only gradual curves (Fig. 10-12). Avoid square or sharp corners in the cutout.

In the case of a hole to be patched that is not damaged around the edges, such as one that was cut out with a saw (for example, a cutout for a window that is to be eliminated and filled in with a fiberglass laminate), it will not be necessary to cut away additional fiberglass provided that the laminate around the cutout is sound. The damaged fiberglass can be cut away with a hacksaw blade in a file-type handle holder or with a saber saw using a fine-tooth blade. If extensive cutting is to be done, use a tungsten carbide blade. These blades are more expensive than standard blades, but they will last much longer. This will more than make up for the price difference. In some cases only minor saw cutting will be necessary, or perhaps none at all. In other cases a cutout all the way around the damaged area will be necessary. Another possibility is to use a power disk sander with coarse sandpaper and to sand off the damaged, weakened, and delaminated fiberglass.

A large bonding surface is necessary in order to adequately join a new fiberglass laminate to the old one. An edge-to-edge butt joint with square cuts would obviously be inadequate. To give a large bonding surface, the area surrounding the cutout or hole needs to be tapered. A single taper from the back side that angles outward to the finished

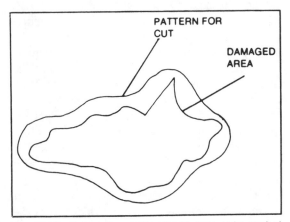

Fig. 10-12. Cutout for repair should have only gradual curves.

side of the laminate is used in most cases. An alternate method is to taper the laminate on both sides to the center of the molding. This method is usually only necessary when the patch must be flush and finished on both sides of the molding, and no overlapping laminate can be added to the back side. An overlapping reinforcement on the back side can add greatly to the structural strength of the repair. This is especially important for crucial repairs, such as those below the waterline on a boat hull. For these repairs, it can mean cutting away a section of a hull liner to get to the back side of the laminate for fiberglassing the backing laminate in place. An alternate method of doing this from one side of the laminate is detailed later in this section.

The distance that the taper should extend back from the cutout or hole depends on the size of the cutout or hole and the thickness of the laminate. An inch taper is about right for holes up to a few inches in diameter or equivalent cutouts on laminates up to about ⅜-inch thickness. Thicker laminates should taper away further. A 4-inch diameter or equivalent repair area should taper about 2 inches. Larger damaged areas, especially on thick laminates, might extend 6 inches or more away from the cutout. Remember that a large taper means a larger patch area that must be sanded and refinished. Less taper is generally compensated for by using a large overlapping reinforcing laminate behind the patch. The laminate can be tapered by using hand surfacing tools, at least to an extent. This is practical for small repair jobs, but a power disk sander is almost essential if large areas are to be tapered in a reasonable amount of time.

When the damage does not extend all the way through the fiberglass molding, or there is some other solid material immediately behind the molding such as wood, concrete, or metal, the repair will be made from one side. No backing form is needed for laminating fiberglass in place. Cut pieces of mat progressively larger, as they will go outward in the damaged area in order to fit the taper. Cloth and other reinforcing material can also be used in the laminate.

Open holes and cutouts are repaired similarly, except that a backing piece is placed behind the area. First consider a repair where you can fiberglass on both sides of the laminate. The usual method is to position the backing flush with the back of the hole, fiberglass a patch in place, remove the backing, and then fiberglass a backing in place. The backing can be of heavy cardboard, wood, metal, or other material. Some materials can be curved to conform to the shape of curved laminates. For compound curves, rigid polyurethane foam can be shaped for use as a backing. Cellophane can be sandwiched between the back side of the hole or cutout and the backing material, so the resin will not bond to the backing. Remove the backing after the resin has cured. Hold the backing firmly in place with tape, props, or clamps. Cut pieces of mat and/or other reinforcing material for laminating the patch in place.

Installing a Backing Laminate

If you can only get to one side of the molding, you can install the backing through the hole. There might also be situations where you can get to the other side of the laminate to apply a backing, but it is impractical to fiberglass on that side. In many cases it will not be possible to remove the backing after the patch has been put in place. This usually isn't a problem, as the backing will not hurt anything if it is just left there. This doesn't allow fiberglassing a backing laminate in place later. You can do without this in many cases, or you can install a backing laminate through the hole or cutout.

To do this, first prepare the surface behind the laminate surrounding the hole or cutout for bonding. You might be able to do this through the hole or cutout by hand or even power sanding. If not, at least wash the area with a clean cloth saturated with acetone. Cut a backing piece from heavy cardboard. This should be the size and shape of the desired overlapping backing laminate. Attach small diameter wires to the cardboard (Fig. 10-13). These wires should pass through one or more layers of dry fiberglass mat (Fig. 10-14). Then thoroughly saturate the mat with catalyzed polyester resin. Give yourself plenty of working time. Work the cardboard and saturated mat through the hole or cutout. Pull it back into correct position by

the wires. Place a wood stick across the hole or cutout. Twist the wires together around the stick to hold the cardboard and mat in position until the resin cures.

Cut pieces of mat and/or other reinforcing material for laminating the patch in place. If wires were used for bonding the mat backing in place, cut them off flush with the surface of the mat after the resin has cured. If no backing laminate is to be used, hold the backing in position with wires in a similar manner, except do not cut off the wires until at least part of the patch has been laminated in place and allowed to cure. If possible, however, cut the wires off, so they will not extend to the surface of the finished patch.

Laminating the Patch

Laminate the patch in place. This can be done in one or more operations. Usually not more than three layers of reinforcing material are applied in one operation. These are allowed to cure before adding the next section, and so on. After the patch is complete (it should extend slightly beyond the original laminate), allow it to cure. If laminating resin was used, apply a thin layer of finishing resin so that the surface will cure tack-free for sanding.

If you can get to the back side, remove the cardboard backing or other material. If a fiberglass backing laminate is to be used, prepare the surface for bonding by coarse sanding and cleaning with acetone. If possible, the backing laminate should extend 3 inches or more beyond the cutout that was made for the repair. Cut the pieces of reinforcing material to progressively larger sizes, so the backing laminate will taper to the original laminate. The piece that is laminated in place last will be the largest. This will result in a gradual taper, which will spread the load over a fairly large area without having a high stress point from an abrupt change in laminate thickness. Fiberglass mat and/or other reinforcing material can be used. Cloth and/or woven roving are often used in combination with mat when maximum reinforcement is required. The thickness of the backing laminate can vary, but usually it is from one-quarter the thickness of the laminate being repaired to approximately the same

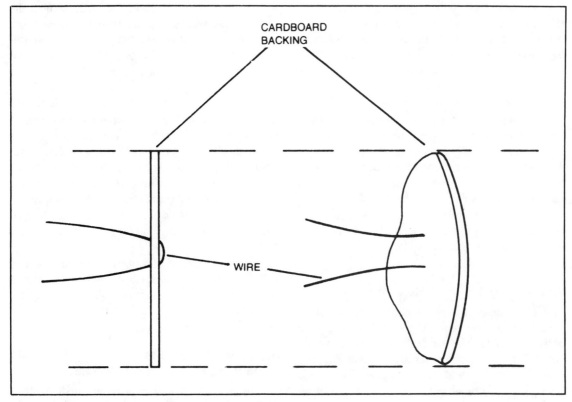

Fig. 10-13. Wire the attachment to the cardboard backing.

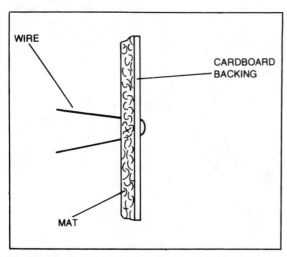

Fig. 10-14. Dry layer of fiberglass mat is fitted to the backing.

thickness. Mat moldings can usually be satisfactorily backed with mat laminates. If cloth and/or woven roving were used in the original molding being repaired, these materials should probably also be used for the backing laminate.

Catalyze polyester resin and apply backing laminate. This may be done in one or more operations. Usually not more than three layers of reinforcing material are applied in one operation. Allow these to cure before adding the next section, and so on. After the backing is complete, allow it to cure.

Only one side of the molding will require finishing in most cases. If both sides require finishing, lay up the patch in two sections. First, taper the edges of the hole or cutout on both sides of the molding to the center of the laminate. Apply cardboard backing or other material covered with cellophane to one side so that it joins the center of the taper. Then laminate half of the patch in place. Allow this to cure. Remove the backing material. The first half of the patch acts as backing for the second half. Laminate the second half of the patch in place. Allow this to cure.

Regardless of the method used for laminating in the patch, sand off excess material on the sides that require finishing. If done by hand use a surfacing tool. A power disk sander will work even better in most cases. When most of the excess material has been removed, work down to finer (higher grit number) sandpaper.

Touch up the gel coat or paint as required on one or both sides of the laminate. Match the original finish as closely as possible.

Constructing a Female Mold Backing

This method is basically laminating a repair in place over a male mold or form. This is usually fairly easy to do. The main difficulty is that you do not end up with a finished surface like that achieved against a female mold. Considerable sanding and fairing is usually required to achieve or approach an against-a-mold type of appearance. In spite of this problem, this method is generally recommended for most typical repairs.

An alternate method that you might want to try after you have had considerable fiberglassing experience is to construct or improvise a female mold backing that is used on the finished side of the molding being repaired. The molding being repaired is usually tapered outward toward the back side. Release agent is applied to the mold surface. If the original surface has a gel coat, a matching gel coat can be sprayed over the release agent. The molding can be gel coated later after the lay-up is complete, and the backing mold has been removed. The repair is then laminated in place against the mold surface.

There are a number of common problems associated with this method. First, it is difficult to get an accurate female mold surface, especially when curved shapes are involved. Second, repair work is often done under less than ideal conditions, making

it difficult to achieve a molded surface without defects. These will have to be filled and sanded, and in the process, much of the advantage of using this more difficult method is lost. Third, it is almost impossible to achieve a joint that does not show between the original molding and the repair molding on the finished side of the laminate, especially if gel coat was applied to the surface of the mold. A small joint line or crack will usually show on the surface. This can be "veed" out and filled with resin putty, but sanding is again required on the finished side of the molding. It also leaves a weakened area, but this can usually be compensated for by applying an extra thick backing laminate that overlaps the repair area on the back side by several inches or more.

This method of molding in a repair laminate is difficult to execute. It should not be attempted until you have had considerable fiberglassing experience.

Repairing Fractures and Holes in Sandwich Core Laminates

Impact and a variety of other conditions can cause fracturing and rupturing of sandwich core moldings. Repair is generally more difficult than similar damage on a single-skin laminate because there are two fiberglass skins and a core to repair instead of a single fiberglass skin. As a rule, if the damage extends all the way through a sandwich core molding, the repair can be thought of as three separate units: one fiberglass skin, the core, and the other fiberglass skin. Steps for making typical repair are as follows.

Carefully survey the damaged area. Whenever possible, examine the damaged area from both sides. Use visual inspection and sound, such as by light tapping with a small metal hammer, to determine the limits of the damaged laminate. Mark off these limits with chalk or a marking pen.

In most cases broken, weak, and delaminated fiberglass must be removed from both fiberglass skins, and damaged sections of core must be replaced or repaired. Mark off the areas to be removed from each fiberglass skin. These should be

150

on or outside the damage limit marks made earlier. Use gradual curves. The two skins need not follow the same pattern. It's usually better if they don't, as this will locate the bonding areas in different places on the opposite skins.

Remove the damaged fiberglass without cutting into or doing additional damage to the core. A power disk sander can be used. At the same time, taper the laminates for bonding in a repair laminate.

Repair any damage to the core. Fill in damaged areas of wood and plywood cores with polyester or epoxy putty if they are not too extensive. For larger areas, a section of the wood can be chiseled away or otherwise removed. Fit a replacement piece of wood and epoxy glue the replacement wood to the original wood core. Rigid plastic foam and balsa cores can also be filled in with polyester or epoxy putty if they are not too extensively damaged. Fill in larger areas with a putty made by adding microballoons to polyester or epoxy resin to form a lightweight filler material. Still another method is to remove a section of the core and splice in a filler piece of the same material, which can be epoxy glued in place.

The core will then serve as a backing for laminating in the two patches. Laminate these in place one at a time.

If you can only fiberglass on one side of the laminate, cut out the damaged area through both laminates and the core. Then laminate in a patch on the backing laminate through the cutout, as detailed previously for single-skin laminates. Repair the core. The core will then serve as a backing for laminating in a patch on the skin on the finished side of the laminate.

Allow laminates to cure. If polyester laminating resin was used, brush on a thin layer of finishing resin on the surfaces to be sanded, so they will cure tack-free.

Sand off excess material. When most of the excess material has been removed, work down to finer (higher grit number) sandpaper.

Touch up the gel coat or paint as required on one or both skins. Match the original finish as closely as possible.

REPAIRING MAJOR DAMAGE TO FIBERGLASS MOLDINGS

Major damage to large areas of a fiberglass molding can usually be repaired in a similar manner, except that larger and often more complex backing forms are required. Plywood makes an excellent backing for large flat areas or areas with simple curves. In some cases the plywood can be bolted to the laminate. If the plywood is to be left in place, countersink the bolt heads in the fiberglass molding (Fig. 10-15). These can then be filled with

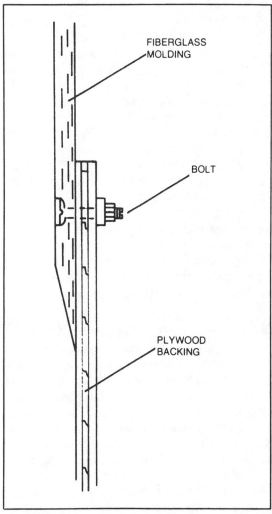

Fig. 10-15. Plywood backing held in place by bolts with heads countersunk in fiberglass molding.

fiberglass putty. If the plywood backing is to be removed, fiberglass the bolt holes in after the bolts have been removed. If the plywood is to be left in place, rough sand it to give a good bonding surface for the first layer of the repair laminate. If the plywood is to be removed, cover it with release agent or add cellophane to the plywood surface so that the laminate will not stick to the plywood. If the laminating must be done at a steep angle or vertical plane or overhead, bonding to the plywood and leaving the plywood in place is generally the best method. This will allow stapling the first layer of the laminate to the plywood.

Complex curves can present additional problems. One method of shaping a backing is to use ribs and strip planking. Another method is to shape rigid polyurethane or polyvinyl chloride foam plastic. Use a surfacing tool to shape polyurethane rigid plastic foam. Thin sheets of polyvinyl chloride rigid foam can be formed to complex shapes over a form when heated. Rigid plastic foam can become part of the laminate or removed after the laminate laid up over it has been completed. If the rigid plastic foam is to be removed, cover it with cellophane before laying up the laminate.

Another method sometimes used for repairing major damage to fiberglass moldings is to mold a repair laminate that will fit the damaged area. Once the repair laminate is formed, it is then fitted to a cutout in the original damaged molding and bonded in place, usually with fiberglass bonding straps.

Molding Repair Laminates over a Male Form

One method of molding repair laminates is over a male form or plug. The male form or plug can be formed from materials like wood, plaster of paris, rigid plastic foam, and metal. The male form or plug can be part of the laminate or not. Rigid plastic foam frequently does become part of the laminate. One method is to shape the foam to the desired finished size and shape, minus the thickness of fiberglass laminate to be placed over the foam. This is similar to laying up a repair laminate in place over rigid plastic foam backing, except that this is done away from the molding being repaired.

A fiberglass laminate is laid up over the rigid plastic foam. Generally, three or more layers of reinforcing material are used. In some cases the other side of the rigid plastic foam is also covered with a fiberglass laminate to form a sandwich core molding.

Many shapes can be formed in this manner. For example, a box-shaped laminate can be formed by first shaping and epoxy gluing sheets of rigid plastic foam together in the desired shape and size of the finished laminate, minus the thickness of the fiberglass skins to be laminated over the plastic foam. The rigid plastic foam can be sawed, filed, and drilled as necessary. Inside corners can be rounded by filling them in with resin putty and using a round wood dowel as a drag to shape and smooth out the putty.

Fiberglass skins are laminated over the rigid plastic foam inside and outside the box and around the edges at the top of the box. The fiberglass laminate is then sanded and finished as desired.

In addition to making repair laminates, many components, such as iceboxes and hatch covers, can be fabricated in a similar manner. Blocks and thick pieces of rigid plastic foam can be shaped as desired and covered all the way around with a fiberglass laminate. Surfboards are frequently fabricated in this manner.

Do not use polyester resin with polystyrene rigid plastic foam because it acts as a solvent to this material. Epoxy resin can be used. Either polyester or epoxy resin can be used with polyurethane and polyvinyl chloride rigid plastic foams.

The main disadvantage of molding repair laminates over a male form or plug is that considerable sanding and fairing are required to achieve a smooth finished surface. Using a female mold can largely eliminate this problem. The main problem is in constructing a female mold of the required repair laminate.

Molds Shaped from Wood

Molds for simple shapes can sometimes be shaped from wood or other available materials. When the female mold is completed, a release agent is brushed or sprayed to the laminating surface so that the resin will not stick to it. The completed laminate can be removed from the mold. A color gel coat can be sprayed inside the mold over the release agent. An alternate method is to skip the gel coat and finish the surface by gel coating or painting after the laminate is completed and removed from the mold. The first layer of the laminate is usually a mat layer, but if desired, cloth reinforcing material can be used. Additional layers of mat and/or other reinforcing material are laid up over this until the desired laminate has been formed.

After the laminate has cured, the molding is removed from the mold. The edges are trimmed as necessary. This can be done by hand with a hacksaw, or a power saber saw can be used.

Several repair components can be molded in a similar manner. You can also make original constructions. There are many possibilities for hobby projects. Stick to simple shapes until you gain the skill and confidence necessary for more difficult projects.

Molds from Flexible Mold Compound

Flexible mold compound is another possibility for making molds. This is useful when you have a small item made of fiberglass or other material that you want to make a mold of for duplication in fiberglass. A release agent is applied to the outside of the object being copied. Successive layers of the flexible mold compound are then brushed or poured on until a mold about ¼ inch thick is formed. This will cure in about 12 hours to a hard rubbery texture, which is then removed from the object being copied. The mold is then used for laminating a fiberglass molding. The mold can be used repeatedly.

This same concept can be applied for making a repair molding, provided that you have a duplicate molding that is undamaged to copy from. Still another possibility is to use available items, such as plastic pans and trays, as molds for laying up fiberglass laminates. Often you will be able to find something that has the shape that you need.

MAIN CONSIDERATIONS FOR REPAIRING FIBERGLASS MOLDINGS

Methods for making many types of repairs to fiberglass moldings have been detailed. The general principle is to replace damaged areas of the moldings with new laminates. Some type of backing, form, or mold is required for laying up the laminate. This holds the resin and reinforcing materials in position until the resin cures. A main problem is in joining the old molding to the new laminate. Proper preparation of the bonding surface is essential. A large bonding surface is important. Whenever possible, patches should be reinforced with backing laminates. Most types of repairs require sanding and fairing. To keep this to a minimum, take care to keep these surfaces as smooth and even as possible during the lay-up process. Whenever possible, methods that give a finished or nearly finished surface should be employed.

REINFORCING AND STIFFENING FIBERGLASS MOLDINGS

Methods and requirements for reinforcing and stiffening fiberglass moldings were detailed in Chapter 9. In fiberglass repair work, it is frequently necessary to make a fiberglass molding or laminate stronger and/or stiffer. There are many reasons for doing this work. The molding might have suffered damage because it was too thin or lacked adequate stiffening. You repair the damage but want to prevent it from happening again. The basic idea is to find the underlying weakness and then to make the necessary corrections. A molding, such as the bottom of a boat, might flex too much when in use. Making the bottom of the boat stronger and/or stiffer can be a worthwhile repair or improvement.

The main methods used for making existing fiberglass moldings stronger and/or stiffer are: thickening the molding by adding to the laminate, adding a backing or core material to the fiberglass molding, and by bonding or otherwise attaching reinforcing members and stiffeners to the fiberglass molding. Weight is added to the molding in each case, which might or might not be important. In some cases, however, weight must be kept to a minimum. The maximum amount of strengthening and/or stiffening must be achieved for the amount of weight added, which must be kept to a minimum.

Thickening Moldings by Adding to the Laminates

One method for strengthening and stiffening an existing fiberglass molding or laminate is to fiberglass in place additional layers to the laminate. In most cases this would have been quite easy to do during the original molding process. It might or might not be easy to add to an existing molding later. For one thing, there can be reinforcing and stiffening members in the way. There can be a liner that has to be removed to even get to the back side of the molding, which is where the laminate is usually added. Another difficulty is that bonding to an old molding can be difficult, especially if it has been painted.

Before deciding to add thickness to a laminate, first determine if it's the best method for accomplishing the desired strengthening and/or stiffening of the molding. It might be easier, and would add less weight, if stiffeners were added instead of thickening the laminate.

If you do decide to thicken a laminate, the usual method is to add the extra laminate to the back side of the molding. In some cases, however, it might be desirable to add this to the finished side of the molding or to both sides.

Before adding the laminate, remove all paint from the bonding area of the molding. This is usually done by sanding. Most paint removers act as a solvent to polyester resin, although there are special paint removers available that are formulated for use on fiberglass.

Rough sand the area to give a good bonding surface. Then thoroughly clean the area using a clean white cloth saturated with acetone. This will remove wax, grease, and oil from the surface.

Decide how thick a laminate is to be added and which reinforcing materials are to be used. The first layer added is usually mat. Additional layers can be mat, cloth, woven roving, or combinations

of these materials. The practice exercises detailed in Chapter 7 should have given you a good idea of the possibilities.

The original molding acts as a mold or form for adding the new laminate. If the laminating can be done in a level or near level plane, use regular viscosity resin. Use high-viscosity resin for angle, vertical, and especially overhead areas. Special high-viscosity resins can be purchased, or thixotropic agent can be added, to regular resin. The less expensive polyester resin can be used in most cases. Sometimes the more expensive epoxy resin is used for bonding the first layer of reinforcing material to the molding. After this has cured, the laminate can then be continued with polyester resin.

In most cases one layer of reinforcing material will be applied and saturated with resin, then allowed to cure before adding another layer. If polyester resin is used, laminating or bonding resin should be used for everything except the final layer of the laminate.

The usual method of application is to first cut reinforcing material to the desired size. Small areas can be covered with a single piece of reinforcing material; larger areas will require sections of material that can be joined by overlapping the pieces of material about 3 inches. Beginners and those with limited fiberglassing experience should not attempt to lay up more than about a square yard of reinforcing material at one time. As experience is gained, larger pieces can sometimes be handled.

Brush on a layer of catalyzed resin to the surface to be bonded to — that is, the bonding surface of the molding or the last layer of laminate added to it. Place the dry reinforcing material in place. One way to do this is to roll it up around a cardboard tube. Then roll it out into position in the wet resin. The reinforcing material is then smoothed out and pressed down into the wet resin. Saturate the reinforcing material with resin. Work out any air bubbles and smooth out the material.

Allow the laminate to cure before adding the next layer. Moldings, stiffeners and other structural members, fiberglass bonding straps and angles, and other fiberglass laminates can be reinforced.

Adding Backing or Core Material to Fiberglass

Sometimes a backing material other than fiberglass can be added to a molding. Often plywood is used. The main problem here is in bonding the plywood or other material to the fiberglass molding. Two methods are frequently used. The first is to epoxy glue the plywood to the fiberglass molding. Proper preparation of both surfaces is important. The second method is to place a layer of mat saturated with wet catalyzed resin on the bonding surface of the molding. Press the plywood in place on the wet mat, clamping or propping it in place until the resin has cured. Mechanical fasteners can sometimes be used, either alone or along with one of the bonding methods.

Another method is to add a sandwich core to a single-skin molding. First prepare the bonding surface of the molding by sanding and cleaning with acetone. Core material can be end-grain balsa or rigid plastic foam. Usually a thin core from ¼ to ½ inch thick is used. The core material can be epoxy glued to the bonding surface of the molding. Allow the epoxy glue to cure.

Laminate a second fiberglass skin over the core material. Usually three or more layers of mat and/or other reinforcing material are used. These often rejoin and taper back into the original molding at the edges of the core material (Fig. 10-16).

Attaching Reinforcing and Stiffening Members to Fiberglass Moldings

Several possibilities were detailed in Chapter 9. These methods can provide considerable reinforcing and/or stiffening without adding much extra weight. More importantly, they are fairly easy to add to existing fiberglass moldings. In most cases attachment can be made by fiberglass bonding alone without the use of mechanical fasteners, although mechanical fasteners can be used in some cases if desired.

Many stiffeners, such as the hat section shown in Fig. 10-17, can be laminated in place over a core material. The core can be wood, rigid plastic foam, cardboard, metal, or other materials. Typical installation of a rigid plastic foam core hat section begins with preparation of bonding surface on the molding where the stiffener is to be installed. The core material is shaped and fitted. While it is not absolutely necessary for the core material to be bonded to the molding, it is a good idea to bond it in place by setting it in wet resin or by epoxy gluing. Cut strips of mat and/or other reinforcing material in progressively wider pieces. Laminate the hat section in place using progressively wider pieces of reinforcing material so that the laminate tapers to the molding being stiffened.

A similar stiffener can be laid up over a wood core. The wood itself, in addition to the fiberglass

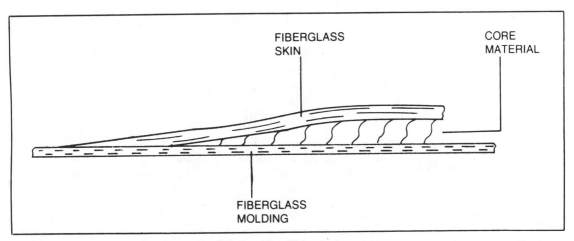

FIBERGLASS SKIN

CORE MATERIAL

FIBERGLASS MOLDING

Fig. 10-16. Core area tapers back into the original molding thickness.

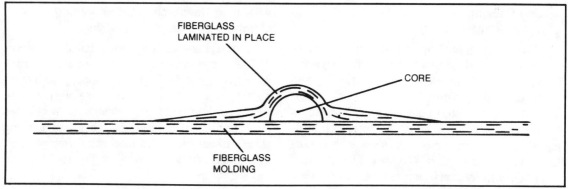

Fig. 10-17. Hat section stiffener.

laminate over it, forms part of the structural strength of the stiffener, whereas rigid plastic foam in the previous example was mainly a form for laminating the fiberglass to the desired shape. It added little structural strength over that of the same laminate with a hollow center. Because it is easier and better to avoid sharp corners when laminating over a form or core, round the outside corners of the wood. Fill and round inside corners with resin putty. Usually the wood is first epoxy glued to the molding or set in place on wet resin saturated mat. When this has cured, add the putty. A round wood dowel can be used as a drag to smooth the putty out. After the putty has hardened, apply the laminate. Begin with a narrow

piece of reinforcing material. Use progressively wider pieces so that the laminate tapers to the molding being stiffened.

It is not necessary to completely encase the wood. An alternate method is to use bonding angle straps (Fig. 10-18).

Still another possibility is to laminate a stiffener in place over a half section of cardboard tubing on plastic pipe (Fig. 10-19). This will leave a hollow center.

There are situations where it is desirable to attach plywood or other materials at right angles to moldings. Plywood bulkheads in boats with fiberglass hull moldings are an example. The joint can be made by laminating fiberglass angle bonding

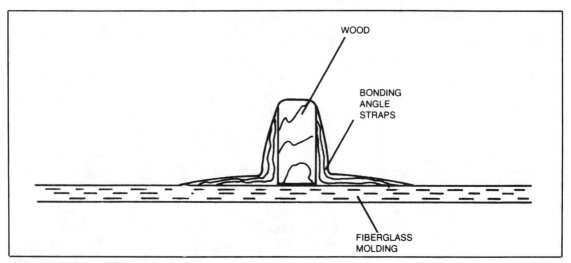

Fig. 10-18. Wood stiffener bonded in place with angle straps.

156

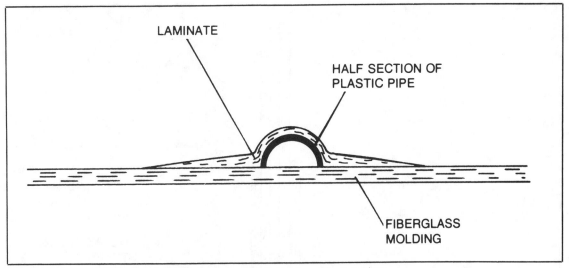

Fig. 10-19. Using a half section of plastic pipe as form for laminating stiffener in place.

straps in place on one or both sides of the plywood (Fig. 10-20). To prevent having a high-stress line where the plywood joins the molding, place a thin strip of rigid plastic foam between the molding and the edge of the plywood. This can also be shaped to give a gradual angle between the plywood and the molding (Fig. 10-21).

A rectangular piece of rigid plastic foam can first be fitted and, if desired, epoxy glued in place. After the epoxy glue has set, shape the rigid plastic foam with a surfacing tool. Cut fiberglass mat and/or other reinforcing material in progressively wider strips. Apply the laminate. The narrowest piece goes on first. Progressively wider pieces then are added. This tapers the bonding laminate to the thickness of the molding and covers the edges of the bonding straps underneath. These methods for attaching stiffeners and other members to fiberglass moldings are useful not only for fiberglass repair work, but also for assembling fiberglass boat kits, as detailed in Chapter 14.

REFINISHING FIBERGLASS MOLDINGS WITH GEL COAT OR PAINT

Both gel coat and paint finishes used on fiberglass moldings and laminates can fade or deteriorate to the point where refinishing might be de-

sired. If a large part of a molding or fiberglass structure has been repaired, it might be more practical to refinish the entire surface than to apply new gel coat or paint just to the repaired area.

Considerable equipment and experience are required to satisfactorily apply gel coat resin to existing fiberglass moldings. In many ways this is even more difficult than applying gel coat inside a female mold at the beginning of a molding lay-up in a contact mold. I've only known of a few professional firms that could do a good job of applying gel coat resin to existing moldings, especially to large ones like boats. It is very expensive to have this work done—often several thousand dollars or more for a 30-foot boat.

Fortunately, there are now brushable two-part polyurethane finishes on the market that give a gel coat like finish (Fig. 10-22). Follow the manufacturer's application instructions carefully. Surface preparation is especially important. Sanding is necessary before application of the polyurethane finish in most cases. These finishes should generally be applied in temperatures between 60 and 85 degrees Fahrenheit. Do outdoor applications in the shade on windless days. The polyurethane coating can be applied with natural bristle brushes and short nap rollers. Many colors are available, and 1 gallon will cover approximately 400 to 600 square

157

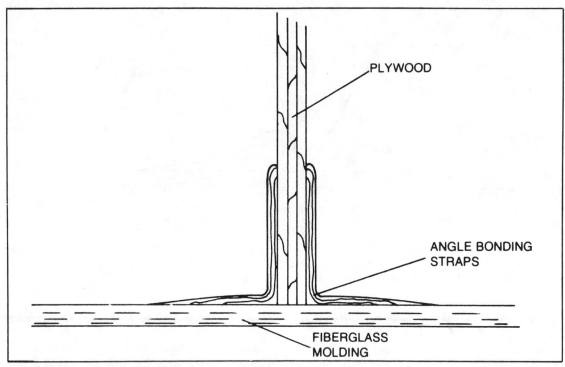

Fig. 10-20. Bonding plywood bulkhead to fiberglass molding.

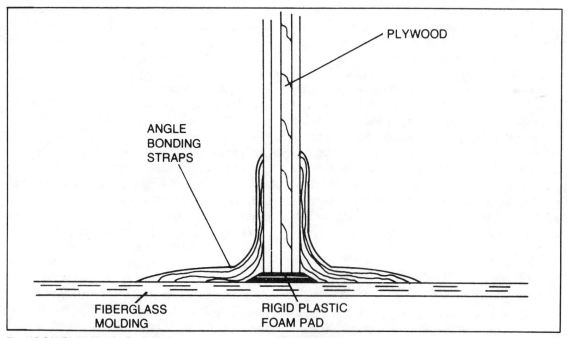

Fig. 10-21. Rigid plastic foam pad used between plywood and fiberglass molding.

Fig. 10-22. Pettit two-part brushable polyurethane paint. (Courtesy of Pettit Paint Company, Inc.)

feet. In most cases the color part is mixed with the hardener part in a 4:1 ratio. For optimum results, two coats are recommended, with about 24 hours waiting time between coats.

Some manufacturers who formerly used gel coats on their moldings have switched to polyurethane coatings, which are applied after the moldings have been removed from the molds. This can give a finish even more durable than gel coating without many of the problems associated with gel coat applications.

Another good possibility is two-part epoxy paints. These have an excellent adhesion factor and give a durable finish. They are generally more diffi-

cult to apply than are the new brushable two-part polyurethane coatings.

Many other paints can be applied to fiberglass. Paint manufacturers and dealers can supply details on the suitability of a particular paint for a particular application.

PRACTICE REPAIR WORK

A major problem faced by most people just starting out in fiberglass repair work is gaining the necessary skill and experience before tackling an actual repair job. The practice projects detailed in Chapter 7 will help toward providing the necessary skill and experience. You can then advance to making repairs on scrap fiberglass moldings, even if you have to create the damage. For example, bash a hole in a scrap molding and repair it. Do the sanding and refinishing.

Fiberglass molding shops frequently have old fiberglass moldings or parts of moldings that they will either let you have or sell to you at low cost. These are ideal for practice work. The fiberglass panels sold at building supply stores are generally of too low quality and too thin to be of much value as practice materials.

Start with simple repair jobs and work up gradually to more difficult ones. The first repairs to actual fiberglass products should be done on items that don't have much value. One good starting place is to purchase a damaged or neglected fiberglass product, such as a pickup shell or boat, and then repair the damage and restore it.

Chapter 11

Repairing Other
Materials with Fiberglass

Fiberglass can be used to repair damage to other materials besides itself. because of the difficulty of bonding polyester resin to most materials besides fiberglass, epoxy resin is usually used when making repairs to other materials. The practice exercises in Chapter 7 gave an opportunity to compare the bonding ability of polyester and epoxy resin to various materials.

Before repairing damage to other materials with fiberglass, see if there is a better, easier, or less costly way of making the repair. With the present high cost of epoxy resin and other fiberglassing materials, it can sometimes cost more to repair something than the item is worth or than it would cost to buy a new one.

For many materials, there is a traditional way of making repairs and a "chemical" way, which often translates to the "epoxy" way. Epoxy has often been called the "repair anything material." There are limits, though. It's important to learn what epoxy can't do, as well as what it can.

REPAIRING DAMAGE TO WOOD

Most damage to wood is best repaired by traditional woodworking methods. There are a few exceptions. Our concern here is with repairs other than covering or sheathing wood with fiberglass, which is covered in the next chapter.

Using Epoxy Putty

While our concern in this book is with fiberglassing and not gluing, epoxy glue is now widely used for joining wood together. Epoxy putty makes an excellent filler material for repairing small defects, holes, and other similar damage.

Before applying epoxy putty, first remove the paint from the wood in the area where the epoxy putty is to be applied. Rough sand the area to give a good bonding surface.

After preparing the surface, measure out and add curing agent or hardener to the epoxy putty according to the manufacturer's directions. Mix together. Use a flat board or other similar object as a mixing surface and a putty knife for mixing. Apply the putty to fill the damaged area with the putty knife. Work the putty into the damaged area and use the putty knife as a drag to smooth and level the surface with the surrounding wood (Fig. 11-1). Epoxy putty generally doesn't have much shrinkage during curing, so the repair area need extend only slightly above the surrounding wood. In fact, too much excess means extra sanding and more waste of epoxy putty.

Allow the epoxy putty to cure, then sand away excess putty. Use a sanding block and sandpaper with no coarser grit than is necessary to take off the excess material in a reasonable amount of time. When approaching the level of the wood surface, take care not to put sanding scratches in the wood. Make certain that the sanding block is held level so that only epoxy putty is sanded away. The epoxy is usually much harder to sand than the wood.

When most of the excess material has been removed, change to a finer (higher grit number)

Fig. 11-1. Applying epoxy putty to wood.

such as one below the waterline on a boat, should have an epoxy resin and fiberglass mat backing laminate to reinforce the epoxy putty fill (Fig. 11-2). You can apply flush backing laminates to one or both sides of the wood by making depressions in the wood (Fig. 11-3). Laminate mat in place with epoxy resin to fill the depressions. After the epoxy resin has cured, sand off excess material to a smooth surface that is flush with the surface of the wood. When done on both sides of the wood, the laminates combine with the putty to form a unit that cannot easily be broken away from the wood.

Repairing Large Holes in Wood

Larger holes and damaged areas that extend through wood and plywood can usually be most easily repaired by adding a wood backing and splicing in a wood patch. There are situations where a

sandpaper. Progress gradually to finer grits until the desired surface is achieved.

The area is then ready for painting. If varnish or other clear finishes are to be applied, add sawdust from matching wood to clear epoxy putty or, for filling small areas, with clear high-viscosity epoxy resin.

Less expensive polyester putty can sometimes be used as a filler for softwoods. The bonding is generally not as good as that of epoxy putty. For hardwoods, used epoxy.

Epoxy putty is commonly available in clear, white, gray, and black, but sometimes wood colors, such as mahogany, are obtainable. These would generally only be useful when large amount of filling are to be done and a natural finish is to be applied, such as varnish or clear plastic urethane. Oils and stains generally won't work over epoxy.

There are many less expensive filler materials for wood on the market. These are also easier to apply than epoxy and work well for many repair applications. Epoxy gives a better bond and stronger repair, however, and it should be used for more crucial repairs.

Epoxy putty alone is usually adequate for filling small, minor holes in wood. Small crucial holes,

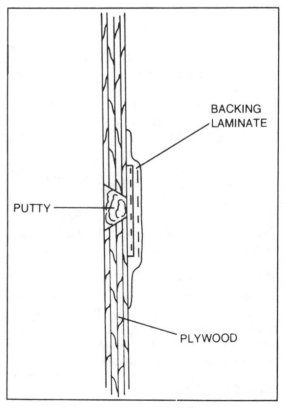

Fig. 11-2. Backing laminate used to reinforce putty fill of hole.

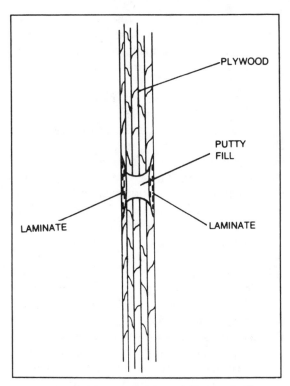

Fig. 11-3. Flush laminates used to reinforce putty fill.

sander. Extend the taper back from the hole or cutout 1 inch or more.

If a backing is to be laminated in place later place a temporary backing piece of cardboard, wood, or other material flush with the wood surface behind the hole or cutout. Sandwich a piece of cellophane between the wood being repaired and the backing. An alternate method is to install a permanent plywood backing at this time, which can be epoxy glued and/or mechanically fastened in place. If the repair is to be laminated flush on both sides, place the temporary backing halfway through the hole or cutout.

Cut layers of fiberglass mat and/or other reinforcing material to laminate all or half the patch, depending on which method is being used. Not

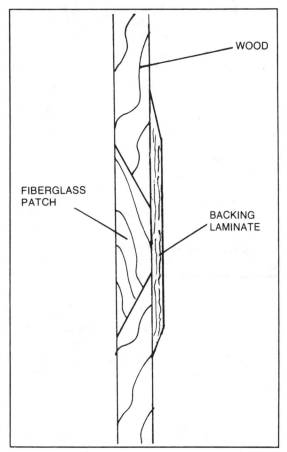

Fig. 11-4. Taper of wood for patching when backing laminate can be used.

fiberglass repair might be desirable. An example is for repairing plywood when the repair cannot extend beyond the surface of the wood on either side and for repair molded wood, such as a molded wood boat hull.

The basic repair technique is similar to repairing damage to fiberglass moldings, except that epoxy resin should be used when repairing wood. Steps for making a typical repair are as follows.

Carefully survey the damaged area. Remove weak, broken, and damaged wood so that bonding of repair laminate will be to sound wood. A cutout will need to be made in some cases. This should have only gradual curves. Mark the area to be cut away. Use a saber saw for cutting. Taper the edges of the cutout to give a large bonding area. Figure 11-4 shows a typical taper when a backing laminate can be applied. If both sides must be finished flush with the surrounding wood, the tapers shown in Fig. 11-5 can be used. Make the taper with an angle saw cut or by power sanding with a disk

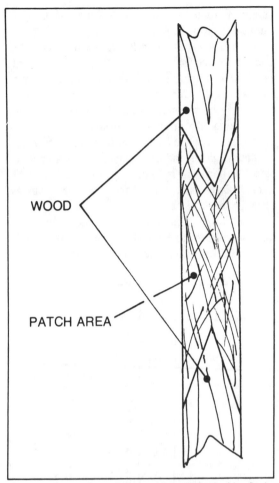

WOOD

PATCH AREA

Fig. 11-5. Taper of wood for patching with both sides flush.

more than three layers of reinforcing material are usually laminated in place in one operation. These are allowed to cure before adding additional layers. This method to cure before adding additional layers. This method should be followed here, at least until you are certain that you can handle more layers. The long curing time of most epoxy resins can slow the work, but a heat lamp or other flameless heating device can reduce the curing time considerably.

Measure out the desired quantity of epoxy resin and add necessary curing agent or hardener to it. Mix well. Apply the first lay-up, which will usually be from one to three layers of reinforcing material. Make certain the reinforcing material is thoroughly saturated with resin. After the first lay-up has cured, add the next section. Continue until the laminate is complete, or half the repair is complete in the case of the laminate that is to be flush on both sides.

If temporary backing was used, remove it. If the entire hole or cutout was laminated in, laminate a backing in place behind the repair area. This should overlap the wood all the way around 3 inches or more. Three or more layers of mat are used in most cases. If desired, cloth and/or woven roving can also be used, either alone or in combination with mat. If half the hole or cutout was filled in, use the first half as a backing for the second half. Laminate the second half in place in the same manner as was done for the first half. Allow the laminate to cure.

Remove excess material on one or both sides by sanding. Small areas can be sanded by hand using fairly coarse sandpaper and a sanding block. For large areas, a power disk sander can be used to greatly speed up the work.

When most of the excess material has been removed, work down to finer (higher grit number) sandpaper. When sanding is complete, paint over the patch to match the finish on surrounding wood.

Repairing Cracks and Weak Seams in Wood

Many other types of damage to wood, such as cracks and weak seams, can be repaired in a similar manner. The basic method is to remove damaged wood, taper to give a large bonding area, and laminate a patch in place using epoxy resin, fiberglass mat, and/or other reinforcing material.

Plywood is sometimes bonded together with fiberglass bonding strips using epoxy resin without the use of mechanical fasteners (Fig. 11-6). This is basically the same technique that is used for bonding fiberglass moldings together. Some plywood boats are assembled in this manner without the use of mechanical fasteners. When properly done, the joints have great strength, sometimes considerably more than that of the surrounding plywood. The result is a plywood boat that is essentially one piece.

This same technique can be applied to wood

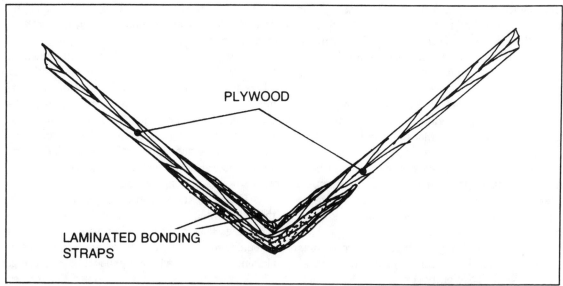

PLYWOOD

LAMINATED BONDING
STRAPS

Fig. 11-6. Method for fiberglass-bonding pieces of plywood together without the use of mechanical fasteners.

repairs. For example, glue and/or mechanically fastened joints that are weak or broken can be repaired in this manner. Remove weak and damaged wood. Taper surfaces to give larger bonding areas. If a finished surface flush with the original wood is desired, remove the thickness of the fiberglass laminate to be added. When the surface is properly prepared, laminate fiberglass mat and/or other reinforcing material in place using epoxy resin.

REPAIRING DAMAGE TO METAL

Most damage to metal is repaired by traditional methods such as welding. Some types of damage can be repaired with fiberglass. This section deals with repairs other than metal auto body repairs and customizing, which are covered in Chapter 13.

While this book is about fiberglassing and not gluing, it is interesting to note that epoxy glue is being increasingly used for joining metals together, especially metal alloys that are difficult to weld.

Epoxy putty can be used for making various minor repairs to most kinds of metal, including aluminum, steel, brass, bronze, and copper. Epoxy putty fillers are available with steel, stainless steel, aluminum, bronze, lead, and other metals in powdered form mixed with epoxy to form a repair

material especially for the metal that it contains. These products even give the finish of the metal. When using these products, carefully follow the manufacturer's directions. Special purpose epoxy fillers are available for repairing cracked engine blocks, cast iron stoves, and other difficult tasks. These fillers sometimes give satisfactory results. Use as directed by the manufacturer.

Repairing Small Dents, Cracks, and Holes in Metal

Minor cracks, small dents, and holes can usually be filled with general-purpose epoxy filler. Before applying epoxy putty, first remove paint, oil, grease, and rust from the metal in the area where the epoxy putty is to be applied. This can be done chemically and/or by sanding or grinding.

After preparing the surface, measure out and add curing agent or hardener to the epoxy putty according to the manufacturer's directions. This can be done on a flat surface such as a piece of wood. Mix the hardener and putty together. Apply the putty to the damaged area with a putty knife. Work the putty into position. Use the putty knife to press it down and smooth and shape the surface. Epoxy putty generally doesn't have much shrinkage during curing, so the repair area need extend

165

only slightly above the desired finished level. Too much excess means more sanding work later.

Allow the epoxy putty to cure. Then sand off excess putty. Use a sanding block and sandpaper with a coarse enough grit to take off the excess material in a reasonable amount of time. When approaching the level of the surrounding metal, take care not to put scratches in the metal. Make certain that the sanding block is held level so that only epoxy putty is sanded away.

When most of the excess material has been removed, change to a finer (higher grit number) sandpaper. Progress gradually to finer abrasive paper until the desired surface is achieved. The area is then ready for painting.

Some types of repair will not require sanding and/or painting. The repair is generally stronger before excess epoxy putty is removed. Removal is usually for cosmetic purposes.

Epoxy fillers can sometimes be used for making small crucial repairs on metal, but use caution here. Avoid repairs that could be dangerous if they should fail. Leave repairs of leaking gas tanks, hot water heaters, and propane tanks to someone specially trained to handle them. Pipes and containers that have liquids or gases under pressure are particularly difficult to repair, and in many cases there are better and/or easier repair methods than by using epoxy filler. If repairs are attempted with epoxy filler, first relieve pressure. Most epoxy will only bond to dry surfaces, although there are epoxies that will bond and cure under water. In most cases drain water tanks and pipes and allow bonding surfaces to thoroughly dry out. Remove paint, rust corrosion, oil, grease, and other chemicals from the bonding surfaces. This can be done by sanding and/or chemical means. Apply epoxy putty. Allow it to thoroughly cure before reapplying water or other pressure to systems. When making repairs on metal with epoxy, it's important not to get epoxy on the threads of fasteners if you want to take them apart later.

Sealing a Leak in an Aluminum Roof

Many repairs can be made to metals by using fiberglass reinforcing material and epoxy resin. For example, leaks in aluminum roofs can be sealed.

Tears land rusted-out areas in sheet metal can be patched. This can be done chemically without welding equipment.

First consider sealing a leak in an aluminum roof. Leaking usually occurs at seams, joints, or where fasteners pass through the aluminum sheeting. Generally, only the area where the leaking is taking place needs be patched or sealed. It is usually a waste of time and money to attempt to sheath a whole aluminum roof or even a large area with a fiberglass laminate, as is sometimes done with wood (see Chapter 12).

The basic steps in making the repair include the following. Locate the area where the leaking is occurring. This might be a seam where two pieces of aluminum are joined or at the edge of a piece of aluminum, perhaps where it joins an edge molding. The leak might be around a fastener that passes through the aluminum.

Decide how large a laminate is required. This should extend several inches or more beyond the limits of the damaged area. Mark off the area.

Remove all paint from the bonding area. Sand the bonding area.

Fill in cracks and low areas with epoxy putty. Build up a smooth cap over the heads of fasteners. The basic idea is to have a bonding surface that is flat or has only gradual curves. Apply the putty with a putty knife. After application, allow it to cure and then sand as necessary.

Cut pieces of mat and/or other reinforcing material for the laminate. While one layer of mat will usually be sufficient to seal leaks, two, three, or even more layers are better. The reinforcing material is usually laid up in progressively larger pieces, but this is less important with aluminum than when repairing fiberglass.

Measure out the necessary amount of epoxy resin and add the curing agent or hardener. Mix together with a stirring stick.

Brush on a layer of epoxy resin to the bonding area of the aluminum and epoxy putty that was applied earlier. Place the first layer of reinforcing material on it. Saturate the reinforcing material with resin. Apply additional layers if needed in a similar manner, either all in one operation or after allowing each layer to first cure. Allow the laminate to thoroughly cure. Sanding and finishing might be required, depending on the particular application. A lacking metal gutter can be repaired in a similar manner.

Patched a Rusted-Out Area in Metal

To patch a rusted-out area in metal, proceed as follows. Sand and/or chemically remove all rust and paint from the area where the laminate will be bonded in place. It is generally best to apply a fiberglass laminate to both sides of the metal, but one side will often suffice. Prepare one or both sides accordingly.

A backing will be needed if there are any holes that pass all the way through the metal. Use cardboard or other material with cellophane sandwiched between the backing and metal, so the laminate will not bond to the backing. The backing can be removed. Backing can be taped, propped, or otherwise held in place.

Cut mat and/or other reinforcing material for laminating the patch in place. All layers should overlap the repair area by several inches or more. If a flush patch is required, use metal body repair techniques detailed in Chapter 13.

Mix epoxy resin and hardener. Laminate the patch in place. A laminate of at least $\frac{1}{8}$-inch thickness will usually be required, either all on one side of the metal or half on each side.

After the laminate is complete or half complete, allow resin to cure. Remove the backing. If half of the laminate is to be applied to the other side of the metal, laminate this in place and allow it to cure.

Sand and finish the laminate as required. If a smooth even finish is required, other laminating methods and surface preparation are required as detailed in Chapter 13.

REPAIRING DAMAGE TO OTHER MATERIALS

Repairs can be made to materials like plastic, ceramic, concrete, and even glass in a similar manner. The main determining factor is usually whether or not the epoxy resin will bond adequately to the material. Before undertaking this type of a repair, first test bond a small piece of fiberglass mat to the material with epoxy resin. If this bonds well, a satisfactory repair might be possible. If you get a poor bond, the repair probably can't be made satisfactorily.

Chapter 12

Covering Wood with Fiberglass

There are situations where it is desirable to cover wood with fiberglass to waterproof it, protect the wood, and possibly improve the appearance. Typical applications are for wood decks, roofs, and boats. Generally, only one side of the wood is covered; the other side is left open so that it can "breathe." A thick laminate can also add to the structural strength of the wood, but this is an expensive method.

Covering wood with fiberglass is a controversial subject. In my opinion, there are certain situations where it can be advantageous and worth the trouble and cost to apply it. In other cases it just doesn't work and isn't worth the time, trouble, and cost.

TYPES OF WOOD SURFACES SUITABLE FOR COVERING WITH FIBERGLASS

Not all wood surfaces are suitable for covering with fiberglass. In most cases only a thin laminate, often with only a single layer of cloth reinforcing material, is used. This provides very little structural reinforcement and is thus suitable only for stable wood surfaces. Plywood surfaces are generally stable, except perhaps at joints, and can usually be covered satisfactorily with a thin layer of fiberglass.

A planked surface such as a picnic table with the top planked with 2 by 4s will probably have expansion and contraction between boards with changes in weather conditions. It will not be a

stable wood surface suitable for covering with fiberglass. A very thick and strong fiberglass laminate might work, but this would be too costly to be practical.

Wood boats are frequently covered with thin layers of fiberglass. Not all boats are suitable or practical for this application. Stable wood boats — boats that act as though they are of one piece — can usually be satisfactorily covered with fiberglass. Examples are plywood planked boats, molded wood boats (molded to shape from thin veneers of wood glued and often also mechanically fastened together to form a structure similar to plywood but with simple and compound curves formed or molded in), and strip planked boats (Fig. 12-1). *Carvel* (Fig. 12-2) and *lapstrake* (Fig. 12-3) are generally unstable and not practical for covering with fiberglass. If they are covered, a very thick laminate is required to resist play and movement between planks from hull twisting and expansion and contraction from changes in weather conditions. There is a method that uses a wire mesh base that can be used for larger boats. This is detailed later in the chapter.

The stability of the wood surface is an extremely important consideration. Covering wood

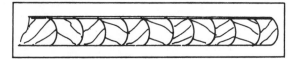

Fig. 12-1. Strip planking

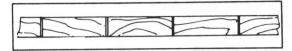

Fig. 12-2. Carvel planking.

with fiberglass requires fairly expensive materials, and the money spent will be wasted if the fiberglass covering does not hold up.

Bonding

Bonding is another important consideration. Polyester resin is generally suitable only for softwoods such as Douglas fir and other woods used for making standard plywood. Bonding with polyester resin can be difficult, and delaminating is a distinct possibility. There are methods that can be used to help assure a reasonably good bond with polyester resin, and they will be detailed later in the chapter.

Bonding is generally better when new wood or plywood is used than when old wood, especially if it has been painted, is to be covered. When attempting to bond to old wood, the condition of the wood is an important consideration. The wood must be sound without dry rot or worms (such as marine teredos), at least no live ones. You might have heard that one of the purposes of covering wood with fiberglass, especially wood boats, is to prevent rot and teredos (shipworms) from getting into the wood. A fiberglass covering might help prevent these conditions, although the covering might actually assist the dry rot fungus in their formation and spread.

Prevention is not the same thing as when the condition already exists. Both dry rot and teredo worms can greatly reduce the structural strength of wood, even if further damage can be halted. In the case of dry rot, this can be a big "if." There are chemicals available for disinfecting wood. These chemicals often contain oils that will make bonding to the wood even more difficult. There are also epoxy base rot cures. These soak into rotted wood and permanently reinforce the wood, but only to the parts of the wood to which you can get the epoxy. The epoxy base rot cures usually come in two-part systems. A curing agent or hardener (one part) must be added to the epoxy (the other part) before it is applied to the wood. This treatment is expensive, usually from about $10 to $25 for a quart (total amount of the two parts) of the special epoxy resin and hardener. It can be effective, especially for small areas of dry rot, and there is usually no problem in bonding fiberglass to this epoxy.

For bonding, the wood should be as dry as possible. Dry rot might more appropriately be called "wet" rot, because moisture and water are necessary for the formation of dry rot. Even if you can somehow dry out wood that has dry rot, it's still difficult to bond to dry rot areas unless they are first treated with the epoxy.

Epoxy resin generally provides a much better bond even on wood surfaces where polyester resin can be used satisfactorily. The better bonding, however, must be weighed against the higher cost. Polyester resin is unsuitable for bonding to hardwoods. Instead, use epoxy resin. Only the first layer of reinforcing material that is bonded directly to the wood need be applied with epoxy resin. After this has cured, the less expensive polyester resin can be used for laminating additional layers in place.

Purpose of Covering the Wood

When considering what types of wood surfaces can be successfully covered with fiberglass,

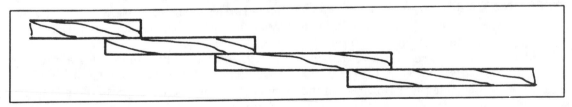

Fig. 12-3. Lapstrake planking.

169

the purpose of covering the wood must be considered. Thin laminates only seal and protect the wood in a manner similar to a paint coating, without adding much structural strength. The possibility of adding a thicker and stronger fiberglass laminate that has considerable structural strength of its own also exists. This becomes a composite wood and fiberglass structure, although how much one enhances the other depends on many factors. Because wood and fiberglass have different physical properties, there can be problems. Almost any wood structure could conceivably be covered with a fiberglass laminate that would be structurally strong enough to hold together. In many cases this laminate would have to be so thick and costly, though, that it would be impractical. Why not just construct it right out of wood or from fiberglass?

Perhaps you have an old and quite large wooden boat that has more or less ended its days as a wooden boat. The boat is beyond the stage where it is practical to repair or recondition it by traditional woodworking methods. Is fiberglass the answer?

I've seen many attempts at reviving just such a boat. Most were more or less unsuccessful — an expensive means of postponing the end of the useful life of a wood boat for a short period of time. The added weight generally made the boats poor performers. There were a few successes, too. These were usually large boats that could take on considerable added weight and still perform satisfactorily. A special method, such as using a wire mesh as detailed later in this chapter, was used. Be forewarned, however, that this type of laminating requires advanced skills and the materials are quite costly. In my opinion, such a project should not be undertaken without the advance approval of a competent naval architect.

TYPES AND THICKNESSES OF LAMINATES

My concern will mainly be with fairly thin laminates with one to three layers of reinforcing material used, except over seams, where additional layers can be used advantageously. The cost of the materials alone will make anything more than this prohibitive. Also, in many cases the added weight

must be kept to a minimum anyway, so only a thin laminate is practical. A few calculations from your notes on the practice exercises given in Chapter 7 will show how quickly various thicknesses and types of laminates can add on weight.

Wood is usually covered with fiberglass cloth, although nonfiberglass polypropylene, polyester, and acrylic cloth reinforcing materials offer many advantages, often at very little additional cost. They also weigh less than fiberglass cloth that will give the same laminate thickness.

Why use cloth rather than mat or woven roving? For its thickness, cloth gives the strongest laminate. It also gives the smoothest surface — that is, it will require less sanding and fairing. Cloth provides a strong, thin laminate. If properly bonded to the wood, the surface of the laminate should closely imitate that of the wood below it. On the other hand, mat tends to bunch up in spots and otherwise give an uneven surface, which will require sanding and fairing to achieve a smooth finished surface. A smooth finished surface isn't always necessary on roofs and various other surfaces. Woven roving presents difficult bonding problems when laminated to wood, although it can and has been done successfully. It also leaves a coarse weave pattern showing on the surface, which is difficult to finish.

While wood surfaces up to a yard or so square can be covered with a single piece of reinforcing material, larger surfaces generally require more than one piece of material for two reasons. It is difficult to lay up larger pieces at one time. Reinforcing materials only come in certain widths, although long lengths (sold by the running yard from rolls) are available. Usually lap joints (Fig. 12-4) are used rather than butt joints (Fig. 12-5). If a perfectly smooth surface is required, the edge of the first piece of material can be feather sanded after the lay-up and cure of the resin. The second piece is then laminated in place with a lap joint. After the laminate cures, the joint is then sanded flush.

If lap joints without this feather sanding are used, cloth will give the least distortion of the surface because of its thinness. Both mat and woven roving will give considerable distortion.

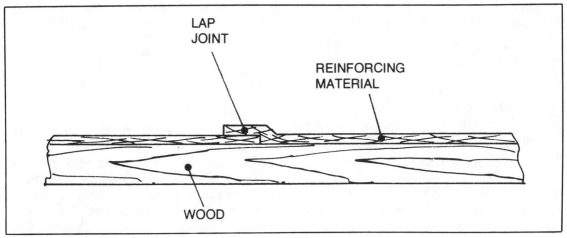

Fig. 12-4. Lap joint.

With the feathering method, all materials will give a flush surface.

If more than one layer of reinforcing material is to be used, butt joints are sometimes used. They are located in different places on each layer. There is usually some structural weakness between pieces of material, but this normally wouldn't be critical when applying waterproofing and/or protective coverings to wood. The edges at the joints do tend to stick up slightly, but minimal sanding is usually all that is required. The sanding should be done after each layer has been applied and allowed to cure, but before the next layer is added. I've used this method with three layers of mat to cover a plywood cockpit floor on a boat with excellent results.

If seams and other areas are to be reinforced, cloth tape (narrow widths of cloth reinforcing material with selvaged edges) can be used. This can be applied without the edges coming unraveled. This gives an even edge and makes feather sanding easier.

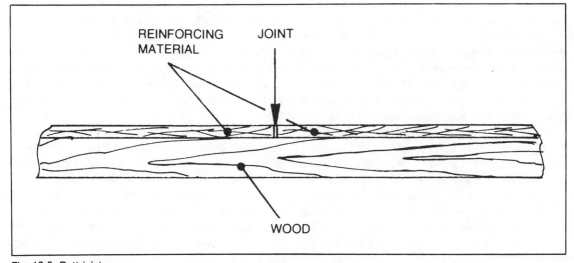

Fig. 12-5. Butt joint.

AMOUNT OF
REINFORCING MATERIAL REQUIRED

After deciding the type(s) of reinforcing material and number of layers to be used, take measurements to determine the amount of material required. Allow for lap joints, overlapping edges, and any areas that require additional layers. It is best to purchase a little extra, as there will likely be some waste and possible mistakes that might ruin some of the material.

AMOUNT OF RESIN REQUIRED

After you have determined the type, weight, and amount of reinforcing material required, the amount of polyester and/or epoxy resin that will be needed can be estimated by using methods given in Chapter 7. It's best to purchase some extra, as there is almost certain to be some waste, such as resin that gels before it can be applied.

PREPARATION OF WOOD SURFACE
FOR COVERING WITH FIBERGLASS

Preparation of wood surfaces for bonding fiberglass to them varies, depending on whether new or old wood is being covered.

New Wood

New wood has the advantage of being in sound condition and not having paint to remove. The wood should be as dry as possible.

Shaping and filling might be required. Because fiberglass bonds best to flat surfaces and gradual curves, avoid sharp corners when possible. Round outside angles off. Round inside angles by adding wood moldings in the corner or use polyester or epoxy putty. After applying the putty, use a round wood dowel as a drag for smoothing and shaping it. Fill holes, depressions, and defects in the wood with polyester or epoxy putty, especially if they are large enough to show in a covering laminate.

Opinions vary on how new wood should be sanded in preparation for fiberglassing, or even whether or not sanding is necessary. As a rule, epoxy resin will bond to new wood without additional sanding provided that the wood is reasonably

dry. Rough sanding won't hurt, though, and might well improve the bonding. I have found that rough sanding with coarse sandpaper to scratch and burr the wood surface improves bonding when polyester resin is used.

Old Wood

It is much more difficult to bond fiberglass to wood that has been in use, especially if paint or other finishes have been applied. Also, moldings and various appendages that you want to fiberglass under are likely to be in place. These should be removed until the fiberglassing is completed. They can then be reinstalled. The fiberglassing might make some of them unnecessary.

Paint can be removed by sanding or by using a paint remover. Use a type that does not contain oils or waxes that can be absorbed by the wood. Another method for removing paint is to burn it off with a blowtorch. A blowtorch should not be used here because it tends to drive oils from the old paint into the wood.

If sanding is to be used for removing paint on large areas of wood, a power sander is almost essential. Pad sanders (orbital or reciprocal action) are generally too slow. Disk and belt sanders work well for removing paint, but take care so that an uneven wood surface does not result. The basic idea is to sand off the paint without disturbing the wood surface below. Oversanding must be especially avoided when sanding fir plywood; this tends to result in a wavy surface.

Remove all grease and oil from the surface. Wash the surface with acetone. Apply by saturating a clean white cloth with acetone and scrubbing the surface. Allow the acetone to dry or evaporate before fiberglassing.

Remove loose putty and filler materials and refill with polyester or epoxy putty. Shaping and filling might be necessary to eliminate sharp corners, as detailed earlier for new wood.

Weak, damaged, and rotted wood should be repaired and/or replaced. Dry rot can sometimes be treated by using epoxy base rot products, as detailed previously. Fiberglassing over active dry rot usually will not halt the spread of dry rot in the wood, and further weakening of the wood structure is likely.

APPLYING FIBERGLASS CLOTH LAMINATES TO FLAT WOOD SURFACES

Thin fiberglass cloth laminates can be applied to stable wood surfaces such as plywood roofs, decks, and floors. You must decide what to do about edges. If you merely end the cloth laminate flush with the edge of the wood surface being covered, water will usually saturate easily into the wood at the edges and can cause the fiberglass to delaminate from the wood.

One solution is to continue the laminate on around the edge of the wood (Fig. 12-6) and sometimes on around to underneath the wood (Fig. 12-7). A trim piece of aluminum, wood, or other material can be applied as shown in Fig. 12-8 to prevent the fiberglass from delaminating at the edge.

Edge moldings are sometimes attached to the edges over the fiberglass, as shown in Fig. 12-9, but leaking around fasteners to the wood below is frequently a problem when this method is used. If the method is used, apply caulking or sealant to the fasteners. This will not always eliminate the problem. Often when I have removed this type of molding from fiberglass-covered wood, I've found dry rot around the fastener holes in the wood beneath the fiberglass. Sometimes the dry rot had spread to large areas of the wood and greatly weakened the wood structure.

It doesn't make much sense to add a waterproof covering over wood and then make holes in the covering. Avoid making holes through the fiberglass covering unless absolutely essential.

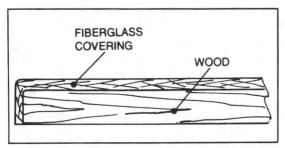

Fig. 12-6. Continuing laminate over the edge of the wood.

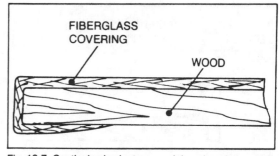

Fig. 12-7. Continuing laminate around the edge of the wood.

Wet Application

There are two basic methods of applying fiberglass cloth laminates: wet and dry. In this method resin is first applied to the wood surface. The fiberglass cloth is positioned on the wet resin, pressed into position, and smoothed out. The upper surface of the cloth is then saturated with resin, until the cloth is fully wetted out and has the proper resin to glass ratio.

This method was detailed in Chapter 7 for adding a 6-inch square piece of 10-ounce-per-square-yard fiberglass cloth to plywood using polyester resin. The same exercise was later repeated using epoxy resin. Essentially the same methods are used here, except that you will be dealing with much larger surface areas.

Polyester Resin Application. Fiberglass cloth is available in weights from 4 ounces or less a square yard to 20 ounces or more a square yard. The light weights usually don't give enough strength or thickness for most wood covering ap-

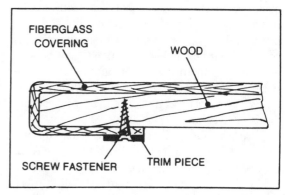

Fig. 12-8. Trim piece is used to keep the edge of the laminate in place.

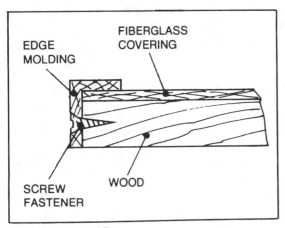

Fig. 12-9. Edge molding.

plications; the heaviest weights are difficult to handle and wet out. The medium weights, from about 7.5 ounces to 12.5 ounces, are most often used. Any weight of fiberglass cloth can be applied with polyester resin by the wet method, although whether or not the bonding to the wood will be adequate depends on those factors discussed previously.

Decide how large the pieces of fiberglass cloth should be. In order to cut down on the number of lap joints that will be required for joining pieces, the pieces should be as large as possible. The pieces should not be so large that they will be difficult to apply, smooth out, and saturate with resin before the resin starts to gel. If a flush surface is required, more separate pieces means more feathering work with a sander. Generally, about a square yard of 10-ounce fiberglass cloth can be handled fairly easily. This would require about 10 ounces of polyester resin—an amount that is practical to handle in one operation (a pint of polyester resin is approximately 16 ounces). As experience is gained, you might be able to handle even larger pieces of fiberglass cloth, up to 2 square yards or even more.

Cut the fiberglass cloth to the required sizes. They can be laid out in a pile with the piece to be applied first on top, the one to be applied second below that, and so on. The material can be rolled around a cardboard tube as desired. Regardless, material should be kept clean and dry. Store material away from the area where you are applying resin until you are ready to apply that piece of material, so you do not splatter resin on the material that is to be used later.

To start the lay-up, you will only need the first piece of fiberglass cloth that is to be applied. Let's assume that you will be using square yard pieces of 10-ounce fiberglass cloth, but other sizes and weights can be applied by varying the amounts of resin.

The decision to use laminating or finishing polyester resin depends on several factors. If the complete laminate is to be only one layer, finishing resin can be used. If feathering is required, this sanding can be done without first applying additional resin so that the surface will cure tack-free. The sanding will also remove the wax in the area where the lap joint of the next piece of material will be bonded.

Laminating resin can be used if more than one layer of cloth laminate is to be applied. This generally creates problems in feathering edges by sanding as the tacky surface of cured laminating resin quickly gums up sandpaper. When laminating resin is used, the final layer of resin over the last layer of laminate should be finishing resin to give a tack-free surface, which can then be sanded.

I generally do this type of fiberglassing with general-purpose or finishing resin. Before additional layers are bonded on, I remove the wax from the surface with acetone applied on a clean white cloth. The resin cures to a tack-free surface for sanding, and I can use one type of resin throughout.

When you are ready to begin the first lay-up, pour out 10 ounces of resin into a mixing container. Add catalyst for about 45 minutes working time. If you later find that this isn't enough working time, reduce the amount of catalyst accordingly. If you find you don't need this much working time, increase the amount of catalyst accordingly.

Mix catalyst into the resin with a stirring stick. Wait about 30 seconds, then brush on an even layer of resin on the bonding surface of the wood for the first piece of fiberglass cloth. Place the dry fiberglass cloth in position on the wet resin. Smooth it out, either by hand (covered by protective gloves) or with a squeegee or laminating roller.

Brush on additional resin to the surface of the fiberglass cloth. Thoroughly saturate the cloth with an even layer of resin. Work out all air bubbles. When everything looks okay, clean the brush with acetone before the resin has a chance to set up. If you mixed too much resin, the excess is usually discarded. Do not apply extra resin to the laminate as the result is a resin-rich laminate that is subject to brittleness and cracking. If you don't have enough resin, quickly catalyze some more and apply before first resin has set up if possible.

Allow the resin to cure. If a flush surface is desired, feather the edge of the first piece for the lap joint, with the second piece to be applied. After sanding, clean the area with acetone on a cloth.

Apply the second section in the same manner. This is then allowed to cure. Additional sections are applied in the same manner until the entire surface is covered. Edge pieces are often fiberglassed around the edge of the wood area, as detailed previously.

If only one layer is to be applied and a flush surface is desired, sand the surface of the laminate as necessary. Extreme care must be taken so as not to sand through the laminate, which is quite thin.

If a second layer of cloth is to be added to the laminate, apply it using the same techniques as for the first layer. The second layer need not be cloth. Mat can be used. Though seldom used for covering wood, the second layer could also be woven roving. Laminating additional layers to the first layer is essentially the same as laying up regular fiberglass laminates. See Chapter 9.

Epoxy Resin Application. Fiberglass cloth can also be applied by the wet method to wood using epoxy resin. The longer curing time generally means that larger pieces of reinforcing material can be used. This means less edges to feather sand if a flush surface is desired. This advantage is minor considering the greater difficulties involved in using epoxy resin instead of polyester resin. You must get the fiberglass cloth to stay in position for a long period of time until the epoxy resin cures. Heat lamps or other flameless heating devices can be used to speed up the cure, but it still takes much longer than polyester resin.

First, cut fiberglass cloth to desired sizes. Because pieces several yards or more in length can be laid up at one time, it is convenient to roll these up on cardboard tubes. After resin is applied to the wood, these pieces can be rolled in place on the wet resin.

Measure out the required amount of epoxy resin. Add curing agent or hardener according to the manufacturer's directions. Mix the resin and curing agent or hardener together with a stirring stick.

Brush on an even layer of resin on the bonding surface of the wood for the first piece of fiberglass cloth. Then place or roll from a cardboard tube the fiberglass cloth in position on the wet resin. Smooth it out. This can be done with glove-covered hands or by using a squeegee or laminating roller.

Although perhaps not essential, at this point I usually staple the fiberglass cloth to the wood using a heavy-duty staple gun and monel staples. I place the staples at approximately 3-inch intervals and make certain that the cloth is stretched tight and smooth. This not only provides a mechanical fastening to the wood but also holds the cloth in position during the long curing time of most epoxy resins.

Some epoxy resin will get on the tip of the staple gun. Clean it off with epoxy solvent before the epoxy has a change to set up. If this is not done, the staple gun will probably be ruined. Carpet tacks can be used in a similar manner, but I prefer the staple method.

Brush epoxy resin onto the fiberglass cloth. Saturate it fully with resin. When everything looks okay, clean the brush with epoxy solvent.

Heat lamps or other flameless heating devices can be used to speed up the cure. Even with heat lamps, it will usually take several hours or more for most types of epoxy resin to cure to the point where it can be sanded. Some types will take much longer.

If a flush surface isn't required, additional sections of cloth reinforcing material can be applied using lap joints without waiting for the first section to cure. A large area can be covered. While it will probably be impractical to use heat lamps or other flameless heating devices, you will only have one long wait for everything applied to cure instead of several shorter waiting periods.

If additional layers of cloth or other reinforcing material are to be added, the less expensive polyester resin can be used if desired. Allow the epoxy resin to fully cure first, however.

A wet lay-up with epoxy resin is generally more difficult than a wet lay-up with polyester resin. The epoxy resin gives a superior bond, however. Epoxy resin is usually much more expensive than polyester resin.

Dry Application

In this method the fiberglass cloth is applied dry to dry wood without a bonding layer of resin being applied first. The resin is then applied to the fiberglass cloth with, hopefully, resin going through the weave of the cloth and bonding the cloth to the wood. The cloth is fully saturated with resin from the one-side application.

This method was detailed in Chapter 7 for adding 6-inch square pieces of 10-ounce-per-square-yard fiberglass cloth to wood using polyester and epoxy resins. Essentially the same methods are used here, except that you will be dealing with much larger surface areas.

For most applications, medium weights — from about 7.5 ounces per square yard to 12.5 ounces per square yard — of fiberglass cloth are used. For dry applications, the cloth should have a large open weave pattern so that the resin can saturate through the material for bonding to the wood.

Larger areas can usually be worked using the dry method instead of the wet method. Often 3 square yards or more of cloth can be wetted out in one operation.

The usual method is to cut all cloth necessary for a one-layer covering of the entire wood surface. These are then placed in position dry on the bonding surface of the wood and stapled or tacked in place with lap joints of about 3 inches where separate pieces of material join. This method will not work if feather sanding is to be used for a flush finished surface, in which case one piece of material is applied at a time. The resin is then applied and allowed to cure. The edges are feather sanded. A second section of fiberglass cloth is applied, and so on.

Polyester Resin Application. Laminating resin can be used, but this creates problems if edges are to be feather sanded. For this type of work, I use general-purpose or finishing resin. If additional layers are to be bonded on, I remove the wax from the surface with acetone applied on a clean white cloth. The resin cures to a tack-free surface for sanding, and I can use one type of resin throughout.

From the data in your notebook for the practice exercises in Chapter 7, calculate the amount of resin required for various weights and sizes of fiberglass cloth. Don't mix more resin than you can comfortably handle during the working time of the resin. This varies, but usually a pint or less is catalyzed at a time. Some experienced fiberglass workers can handle a quart or more at a time. You don't have to catalyze enough resin for the entire job at once. When you run out, mix some more and continue application before the first resin has set up. This method works well when you add catalyst for about 45 minutes working time at the particular working temperature.

With the fiberglass cloth already in place dry on the wood, measure out the first batch of polyester resin into a mixing container. Add the required amount of catalyst. Mix with a stirring stick. Then wait about 30 seconds before starting to apply it.

Resin can be applied with a stiff-bristle brush. This will help to force the resin through the fiberglass cloth for bonding to the wood. An alternate method is to place the catalyzed resin in a paint pan and use a mohair roller for applying it to the fiberglass cloth. A squeegee can then be used for spreading out and smoothing out the resin. When fiberglass cloth is fully saturated with resin, clean the brush or roller in acetone. Allow the laminate to cure.

Epoxy Resin Application. Fiberglass cloth can also be applied by the dry method to wood using epoxy resin. Cut and position dry fiberglass cloth over the bonding surface of the wood in the same manner as given earlier for polyester resin application.

Because of the longer curing times, larger amounts of epoxy resin can usually be mixed at a time than when polyester resin is used. Add curing agent or hardener to the required amount of epoxy resin. Follow manufacturer's directions regarding the amount of hardener to be added for the particular working conditions. Use a stirring stick for mixing resin and hardener together.

Epoxy resin can be applied with a stiff-bristle brush. This will help to force the resin through the fiberglass cloth for bonding to the wood underneath. An alternate method is to place resin (with hardener added) in a paint pan and use a mohair roller for applying it to the fiberglass cloth. A squeegee can then be used for spreading and smoothing out the resin. When fiberglass cloth is fully saturated with resin, clean the brush or roller in epoxy solvent. Allow the laminate to cure with or without the use of heat lamps or other flameless heating devices.

If additional layers of cloth or other reinforcing material are to be added over this, use the less expensive polyester resin. Allow the epoxy resin to fully cure first, however. Conventional wet laminating methods are usually used for adding additional layers.

APPLYING OTHER REINFORCING MATERIALS TO FLAT WOOD SURFACES

The wet and dry application methods can also be used for applying fiberglass mat and woven roving laminates to flat wood surfaces. For roofs and walking surfaces, I frequently use mat instead of or in combination with cloth.

Polypropylene, polyester, acrylic, and other nonfiberglass reinforcing materials (these usually have a cloth weave) can also be applied to flat wood surfaces. Because the materials are more expensive than fiberglass cloth (for the same laminate thickness), they are used only when lighter weight, greater strength, or other special properties of these nonfiberglass reinforcing materials are required for a particular job. These materials are applied by wet or dry methods in the same manner

as fiberglass cloth. When applied dry, these nonfiberglass fabrics provide better bonding and easier resin saturation with either polyester or epoxy resin than fiberglass cloth. Some of the nonfiberglass reinforcing materials are so light, however, that they tend to float in wet resin. This presents problems when laying up the material. A thin layer of resin can be used for wet applications for laying down the fabric, so there is not enough liquid for the material to float on. This can be allowed to cure before subsequent application of additional resin, something that generally won't work with fiberglass reinforcing material. For dry applications, staples or tacks can be used to hold the material against the wood and prevent floating.

COVERING OTHER WOOD SURFACES WITH FIBERGLASS

Covering wood surfaces with fiberglass can be more complicated if the surfaces aren't flat and/or near level. For flat surfaces that are angled, vertical, or even overhead, the wet and dry application methods can be used if high-viscosity polyester or epoxy resins are used. You can add a thixotropic (thickening) agent to regular resins to give a higher viscosity.

Curved surfaces and complicated shapes can also be covered by either wet or dry applications. Avoid sharp angles and corners because it is difficult to apply fiberglassing to them. If possible, round corners and fill inside angles with wood moldings or putty, as detailed previously in this chapter.

A main difficulty in fiberglassing wood with a complicated shape is getting the reinforcing material to bond without having air bubbles or pockets form. Mechanically fastening the reinforcing material in place can help considerably especially when epoxy resin, with its generally longer curing time, is used.

COVERING WOOD BOATS WITH FIBERGLASS

A possible way to give new life to an old wood boat is to cover it with fiberglass. I say "possible" because it is certainly not a complete cure-all.

There are many problems associated with fiberglassing a new wood boat that has been designed and constructed with a fiberglass covering in mind. When fiberglass first became available and popular as a construction material, many boat manufacturers produced plywood planked boats that had fiberglass coverings. These boats were often advertised as being just as good as boats with all-fiberglass laminates. This seldom proved to be the case. Many of these boats ended up with dry rotted wood, delaminated fiberglass coverings, and many other problems—often after a few years of use or less. There are thousands of molded fiberglass boats that are still in sound condition after 20 or more years of use, but very few fiberglass-covered wood boats of this age, including manufactured ones, are still afloat and in use. There were many fiberglass-covered wood boats being manufactured 25 years ago. There are practically none today, although these boats remain popular items for amateurs to construct. The problems are magnified many times when attempts are made to cover old wood boats, especially structurally unsound ones, with fiberglass.

Before undertaking such a repair project, carefully consider what the resulting boat will be like. Covering a small boat with a single layer of fiberglass laminate can be fairly expensive and difficult; similar treatment to a large boat can be a major undertaking and very costly. If the result will be only a short extension of the boat's useful life, the repair probably won't be worth the time, work, and expense.

Another common problem is that the added weight of the fiberglass covering will make the boat a poor performer. Before undertaking such a job, get advice from a competent naval architect or small boat designer to see what the resulting performance of the boat is likely to be.

As mentioned previously, it is generally impractical to fiberglass unstable boat hulls and superstructures, such as planked construction with strips of wood that act separately from each other rather than as a single unit. Carvel and lapstrake planking are particularly bad in this regard. Batten seam (Fig. 12-10) and double planking (Fig. 12-11) are generally somewhat more stable, but they are still impractical for fiberglassing unless a very thick and strong laminate is used. I believe that those boats are best repaired as wood boats by traditional woodworking repair methods. A workable method for fiberglassing large unstable boats that can give good results, if the boat can still perform satisfactorily with the added weight, is detailed later in this chapter.

Stable Wood Hulls and Superstructures

We will first consider stable wood structures that act as though they were constructed from a single piece of wood (Fig. 12-12). Included in this group are many plywood planked, strip planked, and molded (plywood) constructions. It's generally more difficult to fiberglass an old wood boat that wasn't fiberglassed originally than one of new wood construction.

The following is the basic technique for covering stable wood boats with fiberglass. If the boat is small enough, turn it upside down and place it on sawhorses or other suitable stands so that the bottom of the hull will be at a convenient working height. If the boat is too big, it might be possible to tilt the boat over on one side for covering half the hull with fiberglass. Tilt it back and over on the other side to fiberglass that side. Fiberglassing the bottom of an upright boat overhead adds to the difficulty, but it can and has been done. It is generally easiest to fiberglass from above onto a level or near level surface. Angles and vertical surfaces are more difficult. Overhead areas are the most difficult. If the bottom of the boat must be fiberglassed from below, the boat should be placed on a cradle

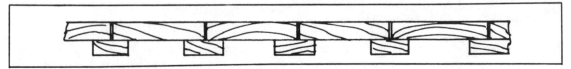

Fig. 12-10. Batten seam planking.

Fig. 12-11. Double planking.

or other type of stand to give working room underneath.

Wood boats are generally covered only on the outside with fiberglass. Sometimes the fiberglass will be used only on the bottom and on hull areas up to just above the waterline. In other cases the entire outside of the hull will be covered. If the boat has a deck, cabin, or other components, these are also often covered with fiberglass. Some wood boats use fiberglass covering only on decks and/or cabin structures, with no fiberglass covering on the hull. There are many possible combinations of areas covered and areas left uncovered.

Wood areas to be covered should be structurally sound and as strong as possible before the fiberglass covering is added. Broken and weak frames, stringers, and planking should be repaired or replaced. Thin fiberglass laminates generally do not add significantly to the strength of the wood structure.

Hardware, rails, and irregularly shaped attachments that are not going to be fiberglassed over should be removed before fiberglassing. If keels, rails, and other wood components are to be fiberglassed over, cove the corners on inside angles and round outside corners for easier fiberglassing and greater strength. Whenever possible, avoid situations where the holes will have to be made through the fiberglass covering for reattachment of wood rails and other similar components. Water will often seep in around fasteners, leading to dry rot in the wood underneath.

Preparation of the surface is extremely important if good or even adequate bonding of the fiberglass to the wood is to be achieved. Remove all loose putty and caulking from seams, cracks, fasteners, and other locations. Replace these with polyester or epoxy putty or, at the very least, with a wood putty that does not contain oil or other substances that will cause bonding problems with

Fig. 12-12. A plywood trimaran being covered with fiberglass.

the fiberglassing resin. Remove all paint, oil, and grease from the wood. Don't use a flame torch, however, as this tends to drive the oils from the paint deeper into the wood. Do not use oil base or wax base paint removers. Use water rinse type paint removers. After using paint remover, sand off any loose scale. Wash the surface with a clean white cloth saturated with acetone. Coarse sanding should be used to give a good bonding surface or "tooth" to the wood. The paint can be removed by sanding without the use of paint remover. A power disk sander generally works well.

Stable boats are most often covered with cloth reinforcing material, although mat and/or woven roving can also be used. Determine how much material will be required for the particular job for one layer; allow for overlaps for joining pieces of reinforcing material. Extra layers may be required to reinforce critical seams, keels, transom areas, and other areas. Decide on the weight, type, and number of layers of reinforcing material to be used. Small boats up to about 20 feet long are generally covered with one or two layers of medium weight fiberglass cloth reinforcing fabric. Larger boats should generally have two or more layers, and heavier weight cloth can be used.

Consider the use of polypropylene, polyester, acrylic, or other nonfiberglass reinforcing material instead of fiberglass reinforcing material. These fabrics usually offer better adhesion and flexibility. They have an open weave that allows dry application, as detailed previously in this chapter. These materials shape better to curved surfaces than fiberglass. Another important advantage is that they weigh less than fiberglass cloth for the same laminate thickness.

Before starting the fiberglassing, have all necessary tools and materials ready and handy. If you are using the wet application method, and first applying a base coat of resin to the wood surface, precut the reinforcing material to the required shapes and sizes. Roll the pieces around cardboard tubes so they can be rolled out over the wet resin base coat.

Mix only small batches of resin at a time. Add catalyst in the recommended amount to polyester resin for desired working time with existing work-

ing temperature. If the resin starts to gel in the container, discard the remaining resin and mix up a new batch. Application of gel resin can create a lumpy mess that can be challenging to remove.

For the wet method, apply the base coat of resin (polyester or epoxy) to the wood surface. While the resin is still wet, unroll the precut cloth. Use a laminating roller or squeegee to press the cloth flat and work out wrinkles and bubbles as the cloth is unrolled.

Without waiting for the first coat of resin to cure, apply a second coat to the cloth. Saturate it thoroughly. Use a laminating roller or squeegee to work out any air bubbles that form. This is the first coat of resin in dry applications.

Allow the resin to cure, then trim edges as required. A hacksaw can be used.

Add additional layers to the laminate as required. If polyester laminating resin is used, coat the final surface with finishing resin, so the laminate will cure with a tack-face surface for sanding.

Sand and fair the surface. Low areas can be filled with polyester or epoxy putty by using a wood strip as a drag. Aid fairing by adding a color pigment, so the putty will contrast with fiberglass covering.

Apply gel coat or paint finish. Two-part brushable polyurethane finish is highly recommended. See Chapter 10.

Decks, cockpit and cabin floors, and other walking surfaces on wooden boats are frequently plywood. These surfaces can often be covered with fiberglass. I have had good results using fiberglass mat. Three layers of 1½-ounce mat works well in most cases. After the laminate has cured, sand the surface. A nonskid paint can then be applied. I often use epoxy paint with a nonskid additive such as *pumice,* a very fine sand.

Unstable Wood Hulls

As previously mentioned, there are many problems associated with attempting to cover unstable wood boat hulls, such as carvel and lapstrake planking, with fiberglass in the manner described for stable wood hulls, even if several layers of reinforcing material are used in the laminate. If a laminate is added that is thick and strong enough to withstand the expansion, contraction, and other stresses from the unstable planking, it is usually thick and strong enough to be a fiberglass hull laminate by itself. What you essentially end up with is a wood boat with a fiberglass hull outside the wood one, with the two hulls not really working together very well. The fiberglass hull is a lot of extra weight, usually enough to make the boat a poor performer. If you are going to go to this much trouble, expense, and hard work, why not build a fiberglass boat using one of the one-off methods? It isn't much more difficult, although be forewarned that both the fiberglass covering job and the one-off boatbuilding require advanced skills, which are beyond the scope of this book.

There is a new and quite different method that has been used to successfully cover a number of wooden boats with fiberglass. Two products not previously mentioned in this book are required: STR-R-ETCH MESH and FER-A-LITE. Both products are available from Aladdin Products, Inc., RFD 2, Wiscasset, ME 04578.

STR-R-ETCH MESH is a special wire mesh with approximately ½-inch square holes between the wire weave. It comes in sheet material 4 feet wide in any desired length up to 500 feet and is rolled for shipping. The material is stiff enough to be self-fairing and can easily be formed into compound curves. FER-A-LITE refers to a system based upon adding a filling called REINFOR-CEMENT to boatbuilding grade polyester resin to form a troweling mixture for use with STR-R-ETCH MESH.

The technique for covering a wood boat is basically to first sand off all paint from the hull and other areas to be covered. Prime the bare wood with a coat of epoxy. After the epoxy has cured, attach wire mesh. Once the mesh is hung in place with a few staples, nails can be used for fasteners. These should be of the same material as the hull fasteners. Drive the nails right next to the wires, so the nail heads will overlap the wires. A quilted pattern on 2-inch centers is suggested for the fastenings.

FER-A-LITE is then used to fill the mesh. This material is light enough to float in water and weighs only 60 pounds per cubic foot. It has a modulus of rupture (bending strength) after curing of 3,600 (pounds per square inch).

To form the mixture, mix REINFOR-CEMENT with polyester resin to form a troweling mixture with a consistency about like mashed potatoes. Add catalyst and trowel the mixture in place to fill and cover the mesh. The light weight of this material makes applying it fairly easy physical work. The material also has high secondary bonding strength, which means that you don't have to apply it all at once. A new batch will bond to that previously applied even after it has cured, which means that the plastering can be done in convenient increments. Curing time depends on the amount of catalyst added, the working temperature, and other factors, but it generally is accomplished in a few hours after the catalyst is added. The surface should be faired as smoothly as possible during the troweling.

After the FER-A-LITE has cured, sand the surface. Then add a fiberglass covering or overlay over it. An alternate method is to press fiberglass cloth onto the surface of wet FER-A-LITE. Then saturate the fiberglass cloth with polyester resin to form a fiberglass skin. Polypropylene or acrylic reinforcing material can be used instead of fiberglass reinforcing material. Sand and finish the surface in the same manner of fiberglass repair laminates.

This technique is fairly inexpensive and very reliable. The mesh gives a strong reinforcement to the wood, yet the total weight of the covering is much less than that of a regular fiberglass laminate with equivalent strength.

Chapter 13

Auto Body Repair
and Customizing

Fiberglassing is becoming an increasingly popular method for repairing and customizing metal automobile bodies. This method is extremely controversial due to improper application. Used cars are routinely repaired in this manner prior to placement on the sales lots. This work is often done with little thought to making a lasting repair. The result is that the fiberglassing repairs often later crack or fall out. The more traditional body solder (an alloy of lead and tin) often also cracks or falls out when it is improperly applied.

When properly applied, quality and lasting repairs can be made by fiberglassing. Only a minimum of tools and equipment are required. No welding equipment or expensive metal straightening tools are needed.

Fiberglassing materials formulated especially for metal auto body repair and customizing have greatly improved over the years that they have been in use. The new materials bond to metal better and are more flexible. Many materials are fairly easy to use.

Many of the leading auto body repair shops are now using fiberglassing techniques to supplement traditional metal straightening and welding repair methods. The trend is toward increased use of fiberglassing methods.

As far as automobile customizing is concerned, fiberglassing methods have largely taken over, not only for the do-it-yourselfer, but also for professional work in commercial shops. Traditional metalworking methods made most customizing dif-

ficult and limited; fiberglassing is generally easier and offers greater creative possibilities.

AUTO BODY REPAIR

Fiberglassing materials and kits for repairing damage to metal auto bodies are readily available at auto supply stores (Fig. 13-1). Several polyester and epoxy filler materials are formulated especially for filling surface depressions and irregularities on metal auto bodies. In the auto body repair field these are often called "plastic" or "cold" filler materials. The term *cold* refers to the fact that a torch isn't required to apply it; there is a heat buildup to cure or harden the putty. These materials are specially formulated so that they will bond well to metal. They have sufficient strength and flexibility to stay in place without cracking or falling out.

These filler materials are usually sold with the necessary catalyst or hardener that can be of liquid, cream, or paste consistency. Use only the specific catalyst or hardener formulated by the manufacturer for the particular type and brand of putty. Do not attempt to make substitutions. When using these materials, carefully follow the manufacturer's directions for mixing and using them. The fillers can be used alone for filling in small areas, or they can be used with fiberglass reinforcing material and a screen for repairing larger areas.

Polyester and epoxy resins are also frequently used for making fiberglassing repairs to metal auto

Fig. 13-1. Bond Tite repair kits (courtesy Oatey Company).

bodies. These are used with fiberglass reinforcing materials and/or a fiberglass screen and sometimes even a metal screen.

A few metal straightening body and fender tools are also useful for preliminary straightening and shaping of metal before fiberglassing (Fig. 13-2). Generally, the most useful tools are the *bumping hammer,* which is used to pound out dents and reshape metal; *dollies,* which come in many shapes and sizes and are used as a backup for bumping hammers; and *dent pullers* (Fig. 13-3).

Also useful are surfacing tools (Fig. 13-4) and power disk sanders. For small jobs, a disk sanding attachment (Fig. 13-5) can be used in a portable electric drill.

Making Repairs with
Polyester and Epoxy Plastic Body Filler

Use brands especially formulated for repairing metal auto bodies and fenders. Carefully follow the manufacturer's directions for proper application. These products generally have a putty consistency. The quantity to be used is measured out, and a catalyst or hardener is added (Fig. 13-6). The putty is then applied and allowed to cure. It can then be filed, sanded, and painted.

There are four basic steps for most applications: surface preparation, mixing, applying, and finishing.

Surface Preparation. Proper preparation of the metal surface is extremely important. Begin by sanding or grinding the damaged surface to bare metal (Fig. 13-7). Remove all paint, rust, oil, and grease. In some cases bumping hammers, dollies, and dent pullers can be used for preliminary straightening of metal to reduce the amount of filler that will be required. Care should be taken, however, so as not to bend or stretch metal to the point where it extends above the desired finished surface level. Break out weak sections of rusted-through metal and concave the edges.

Areas up to about the size of a nickel can usually be filled with filler without additional fiberglass reinforcing material or screen. If a hole goes through the metal, reinforcement is desirable even here. Larger areas should have a reinforcing screen. Cut the screen to fit the depressed area of the metal around the hole. The screen should fit below the desired finished surface area of the filler. Screening can be pressed into wet filler material placed around the hole (Fig. 13-8). Allow putty to cure. The screen then forms the backing and foundation for an additional filler.

Mixing. Follow manufacturer's directions for adding catalyst or hardener. In most cases the mixing can be done with wood applicator or a metal putty knife on a hard, nonporous surface such as glass or metal. The surface should be clean and dry. Thoroughly mix catalyst or hardener with

183

Fig. 13-2. Bond Tite auto body repair tools (courtesy Oatey Company).

filler material. The amount of catalyst or hardener required will depend on working temperature, amount of working time desired, and other factors. Some brands have a system where the filler material changes color when the catalyst or hardener is properly mixed.

Application. Use a putty knife or other applicator for applying the filler material (Fig. 13-9). First, spread a thin layer of filler material over the area. Use sufficient pressure to assure a firm bond to the metal. Then build up the filler to the desired thickness and shape. Slightly overfill the area to

allow for possible shrinkage. Extend the filler approximately 3 inches beyond the depression to allow feather sanding a patch into the contour with the surrounding metal.

Allow the filler material to harden. For most brands and working conditions, the time required will range from about 30 minutes to several hours or more.

Finishing. Finishing begins with removing excess filler material (Fig. 13-10). Use a hand surfacing tool or remove the material by hand or power sanding. Avoid making scratches in the filler

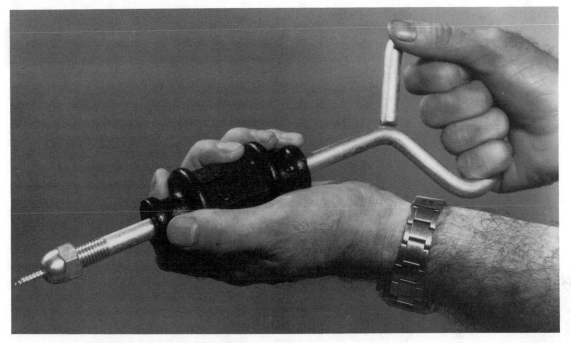

Fig. 13-3. Bond Tite dent puller (courtesy Oatey Company).

material that extend below the desired finished surface level. Use progressively finer grades of sandpaper, usually from 220 grit to 360 grit, to shape and contour the surface and sand to a fine finish (Fig. 13-11).

Fig. 13-4. Bond Tite surfacing tool (courtesy Oatey Company).

If any low spots or scratches extend below the desired finish surface level, mix additional filler and apply to these areas. Allow material to harden, then repeat the finishing procedure described earlier. The basic idea is to blend the patched area into the surrounding surface of the metal. This requires careful sanding.

Minor imperfections such as pinholes or fine sandpaper scratches can be filled with glazing putty. This is similar to a thick primer and requires no catalyst or hardener.

The area is then ready for final sanding. After sanding, prime and paint the repaired area, matching the color of the paint as closely as possible (Figs. 13-12 through 13-14).

Additional Tips. Here are some additional tips for making repairs on metal auto bodies using polyester and epoxy plastic filler. The filler material should not be considered as a substance that will repair anything. Generally, avoid using large quantities of this material without additional reinforcing material. In most cases fiberglass cloth and mat can be used with these filler materials, provided that their viscosity allows saturating the

Fig. 13-5. Bond Tite sandpaper and disk sanding attachment for portable electric drill (courtesy Oatey Company).

reinforcing material. If reinforcing material is to be used, it is better to make the repair with polyester or epoxy resin and fiberglass cloth and/or mat, as detailed later in this chapter.

The filler material should first be mixed in the container in which it is sold with a clean mixing stick. This should be done before the desired amount of filler for the job at hand is taken from the container. Use only the amount of filler that can easily be applied before it starts to gel or harden. Once the filler starts to gel, do not continue the application. Discard this material and mix up a new batch. Do not return any filler that has catalyst or hardener added to the original container, as even a small amount of catalyst or hardener will usually start the curing process. Putting the lid back on the container will not stop the contents from hardening.

Avoid using large amounts of filler in areas subject to bumping and vibration, especially doors, hoods, and trunk lids. If fiberglassing repairs are to

be made in these areas, use the resin and reinforcing material method detailed later in this chapter.

Making Repairs with Polyester and Epoxy Resins and Fiberglass Reinforcing Materials

Use only polyester and epoxy resins especially formulated for metal auto body repair. This is very important in the case of polyester resins, which don't ordinarily bond well to metal. Special polyester resins are now on the market that will bond well. One brand that I'm familiar with is Bond Tite Super Glas polyester resin in gel form (Fig. 13-15). It is manufactured by the Oatey Company, 4700 West 160th Street, Cleveland, OH 44135, and is available from auto supply stores. Because of its gel form, the resin can be applied to vertical surfaces without sag or runoff.

There are five basic steps for most applications: surface preparation, reinforcing material preparation, mixing of resin, application of resin to

Fig. 13-6. Bond Tite auto body repair kit includes filler material and catalyst, applicator, back-up screening, and instructions (courtesy Oatey Company).

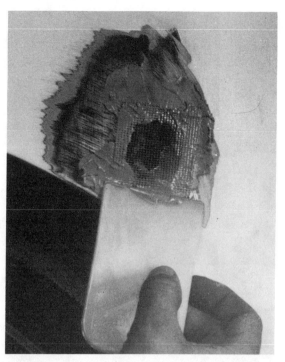

Fig. 13-8. Bond back-up screening in place (courtesy Oatey Company).

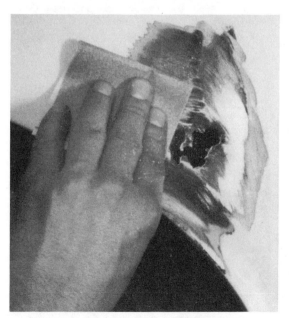

Fig. 13-7. Sand damaged area down to bare metal (courtesy Oatey Company).

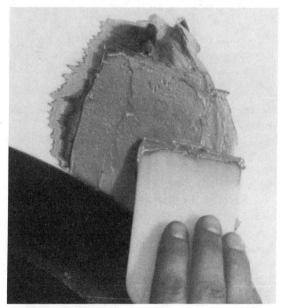

Fig. 13-9. Apply filler material over the back-up screening (courtesy Oatey Company).

187

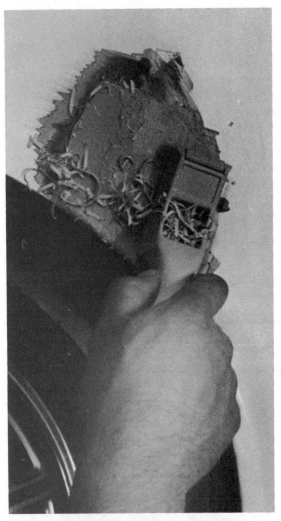

Fig. 13-10. A surfacing tool is used to remove excess filler material (courtesy Oatey Company).

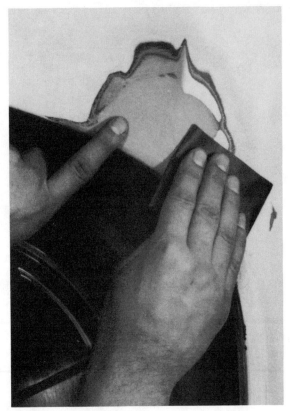

Fig. 13-11. Sand the filler material (courtesy Oatey Company).

metal and reinforcing material, and finishing. See Fig. 13-16.

Surface Preparation. Proper preparation of the metal is extremely important. Begin by sanding. Use coarse grit sandpaper. A power disk sander is ideal (Fig. 13-17). Sand to bare metal, removing all paint and rust. Extend sanding at least 3 inches beyond the damaged area.

In some cases preliminary straightening of metal using a bumping hammer, dollies, dent puller, and other body and fender tools can be

Fig. 13-12. Bond Tite primer in spray cans (courtesy Oatey Company).

Fig. 13-13. Apply primer (courtesy Oatey Company).

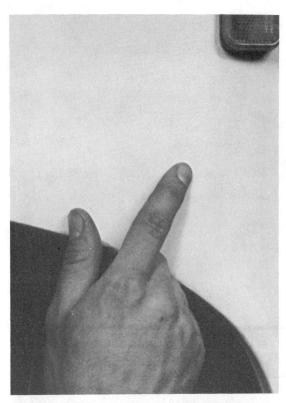

Fig. 13-14. Repaired area after touch-up painting (courtesy Oatey Company).

advantageous. Care should be taken, however, not to bend or stretch metal to the point where it extends above the desired finished surface level. Break out weak sections of rusted-through metal and hammer down the edges to make them concaved below the desired finished surface level. Drill small holes around the rusted-out area to improve resin bonding to the metal.

There should be at least a 1-inch area around any hole to be patched that is below the desired finished surface level for bonding reinforcing material in place. If not, the fiberglass reinforcing material will all be removed in areas around the hole during sanding. Before fiberglassing, the bonding surface of the metal must be clean, rust-free, and without oil or grease present.

Reinforcing Material Preparation. If there is a hole all the way through the metal, some type of backing will be required for laying up the fiberglass patch. If you can get to the back side of the hole, backing can be applied from there. Cut the backing an inch or so larger than the hole all the way around. Use cardboard, metal, or wood. Sandwich a piece of cellophane between the backing material and the back side of the hole. Tape, prop, or otherwise mount the backing in position.

Fig. 13-15. Bond Tite fiberglass repair kit with polyester resin in gel form (courtesy Oatey Company).

Fig. 13-16. Damaged area (courtesy Oatey Company).

If you can get to only one side of the metal, a backing piece can be installed through the hole. Use a cardboard backing. Cut it to a shape an inch or so larger than the hole all the way around. Attach a small diameter wire to the cardboard backing. There is no need for cellophane or release because you won't be able to get to the back side of the metal to remove the backing anyway. Then fold the backing, slip it through the hole, and pull it back in place with the wires. The wires can be attached to a wood prop as shown in Fig. 13-18 to hold the backing in position until the fiberglass patch has been laid up. After part of the laminate is in place, cut off the wires flush. The rest of the

laminate is then added, covering up the ends of the cut-off wires.

An alternate method when you can only get to one side of the metal is to sand and clean the metal around the hole for a distance of several inches on the back side, working through the hole. Take care not to cut yourself on the sharp edges around the hole. Then place a layer of fiberglass mat over the cardboard backing. Trim the mat to the same size and shape as the backing. Pass the wires through the mat (Fig. 13-19). Add catalyst or hardener to the amount of resin required to fully saturate the mat. Saturate the mat with resin. Pass the cardboard through the hole and pull it back into position

Fig. 13-17. Sanding is done with disk attachment in a portable electric drill (courtesy Oatey Company).

with the wires. Use a prop across the hole for clamping the wires (Fig. 13-18). Allow the mat backing laminate to cure, then cut the wires off flush with the surface of the mat. The mat serves not only as a backing, but also reinforces the patch.

Cut fiberglass reinforcing material to shapes and sizes for laminating in a patch. Usually mat and/or cloth is used. The total laminate should extend slightly higher than the surrounding metal after it has cured. From the practice exercises in Chapter 7, you should be able to calculate the

number of layers of various weights of mat or cloth, or a combination of the two, required to give the necessary laminate thickness.

For reasonably small patches that require three layers of reinforcing material or less, the entire lay-up can usually be done at one time with a single batch of resin. Larger and thicker laminates can be laid up in separate layers. Allow the first layer to cure before adding the second layer, and so on.

The top layer of reinforcing material should

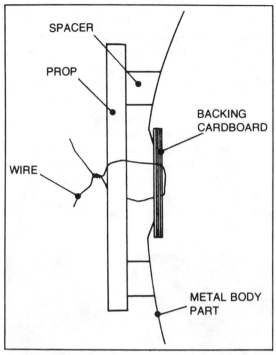

Fig. 13-18. Wires attached to the prop hold backing in place.

extend several inches or more beyond the hole all the way around to allow feather sanding the patch into the contour with the surrounding metal. Organize the dry pieces of reinforcing material so they are close at hand and arranged in a pile. The piece to be applied first is on top, the piece to be applied second is immediately below that, and so on.

Mixing. Estimate the amount of resin required for the first fiberglassing operation. Pour this amount into a mixing cup or container. Add the required amount of catalyst or hardener for the desired working time under the existing working temperature. Follow the manufacturer's directions carefully. Stir the catalyst or hardener thoroughly into the resin using a clean stirring stick.

Application. There are several possible methods for applying the laminate. One way is to brush a layer of resin over the bonding surface. Place the first piece of reinforcing material dry in the wet resin (Fig. 13-20). Position the piece and press it down into the wet resin. Smooth it out. Then brush on additional resin, fully saturating the reinforcing material. Work out any air bubbles or

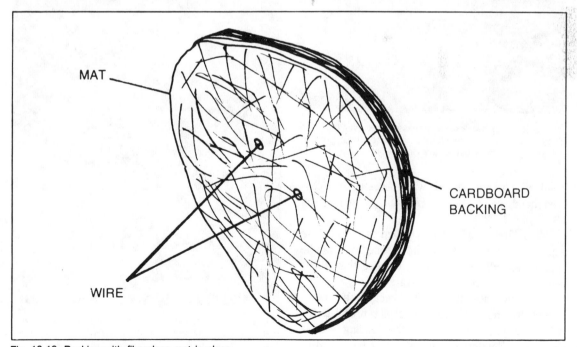

Fig. 13-19. Backing with fiberglass mat in place.

ated by placing it on a piece of cardboard. Saturate one side of the material by brushing on resin. Turn the material over and finish saturating it with resin. Then transfer the reinforcing material from the cardboard to position on the patch area.

Regardless of the method used, it is important to have a bonding layer of wet resin on the metal and to fully saturate the reinforcing material with resin. After the reinforcing material is fully saturated with resin, do not add additional resin. Too little resin can leave dry spots (glass fibers without resin) in the laminate. Too much resin can leave resin-rich areas (resin without glass fiber reinforcement) in the laminate. Both conditions are undesirable and should be avoided.

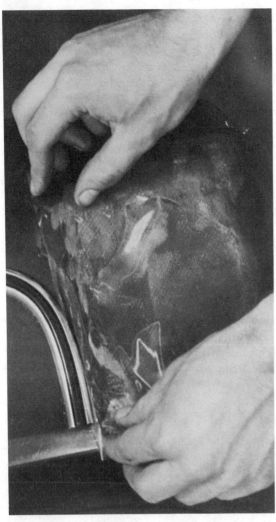

Fig. 13-20. Place wax-coated release paper over the fiberglass reinforcing material that is saturated with gel-type polyester resin (courtesy Oatey Company).

wrinkles. This can be done with a brush, glove-covered hand, squeegee, or plastic spreader (Figs. 13-21 and 13-22).

This first layer can be allowed to cure before applying the next layer, or the second layer can be applied immediately. Usually no more than three layers are applied in one operation.

An alternate method is basically the same, except that the reinforcing material is presaturated with resin before placing it on the bonding layer of wet resin. Reinforcing material can be presatur-

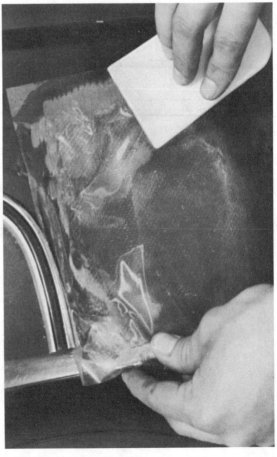

Fig. 13-21. Use a plastic spreader over the wax-coated release paper to smooth out the fiberglass patch (courtesy Oatey Company).

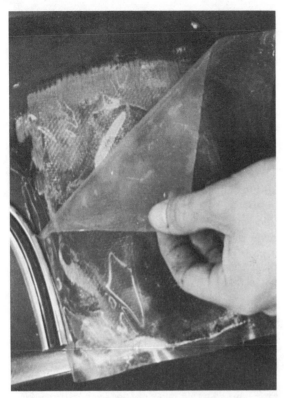

Fig. 13-22. After the resin has cured, peel off the wax-coated release paper (courtesy Oatey Company).

After the patch has been laminated in place, remove the backing cardboard or other material if you can get to the back side of the metal. It might also be desirable to apply a backing laminate that overlaps the metal area on the back side. First, prepare the metal surface for bonding. Sand or grind down to bare metal. Then laminate the backing in place using one or more layers of reinforcing material.

Allow one completed laminate to cure or harden. If epoxy resin is used, heat lamps or other flameless heating devices can be used to speed up curing time.

Finishing. Finishing begins with removing excess material. Use a hand surfacing tool or remove the material by hand or power sanding. Try to avoid making scratches in the patch that extend below the desired finished surface level. The usual method is to use progressively finer grades of sandpaper to shape and contour the surface and achieve a fine finish.

If any low spots or scratches extend below the desired finished surface level, fill these with polyester or epoxy filler, as detailed previously. Allow the filler material to harden, then sand as necessary. The basic idea is to blend the patched area into the surrounding surface of the metal.

Minor imperfections, such as pinholes or tiny scratches, can be filled with glazing putty. This is similar to a thick primer and doesn't require a catalyst or hardener.

The area is then ready for final sanding. Prime and paint the repaired area, matching the color of the paint as closely as possible (Fig. 13-23).

Manufactured Molded Fiberglass Repair and Replacement Components

Many manufacturers (see Appendix for suppliers) are now offering molded fiberglass replacement components for popular metal auto bodies.

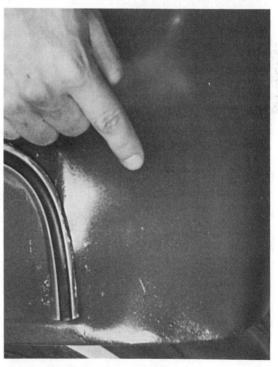

Fig. 13-23. Repair area after painting (courtesy Oatey Company).

Fenders, hoods, trunk lids, tops, and other components are available for selected cars (Fig. 13-24). The molds for most of these were taken from original metal components.

The molded fiberglass components are attached by means by bonding with epoxy glue or fiberglassing and/or mechanical fasteners. For some components, no fiberglassing skills or experience are required for installation; for others, fiberglassing skill and experience are essential.

This method for making repair and replacement parts has made many hard to find metal body and fender components available in fiberglass. When properly installed and finished, it is difficult to tell them from original metal parts. The fiberglass ones won't rust. In some cases molded fiberglass tops are available to replace canvas tops.

CUSTOMIZING

A second major area where fiberglassing materials and techniques can be applied to metal body work is *customizing*. Typical jobs include removing chrome trim, filling in the holes, and reshaping various areas, such as headlights and taillights, to give a distinctive styling. The same basic tools and materials required for body repair are used for customizing work.

Customizing with Polyester and Epoxy Plastic Body Filler

Polyester and epoxy plastic body fillers can be used to fill in small holes and other areas and for reshaping body parts. Use only types of plastic filler formulated especially for application to metal auto bodies. Carefully follow the manufacturer's directions for proper application.

Before applying filler material, first prepare the metal surface for bonding. Remove all paint, rust, oil, and grease from the bonding area. Begin by sanding or grinding the area to bare metal.

Small holes up to about ½ inch in diameter can usually be filled without a screen backing. Taper and concave the edge around the hole (Fig. 13-25). This will give a bonding area for the filler to attach. Larger hole should be prepared in the same manner. A piece of screen is then shaped to cover the hole and extend outward around it in the depression (Fig. 13-26). Catalyze or add hardener to filler material and apply to the bonding area. Press the screen down into the putty below the desired finished surface area of the filler. Allow filler material to cure. A screen then forms the backing and foundation for additional filler.

Carefully follow the manufacturer's directions for adding catalyst or hardener. Use a putty knife or other applicator for applying the filler material. First, spread a thin layer of filler over the area. Use firm pressure with a putty knife to assure metal contact. Then build up the filler to the desired thickness and shape. Extend the filler beyond the depression to allow feather sanding to contour with the surrounding metal. Also, slightly overfill the area to allow for possible shrinkage during curing. Allow the filler material to cure.

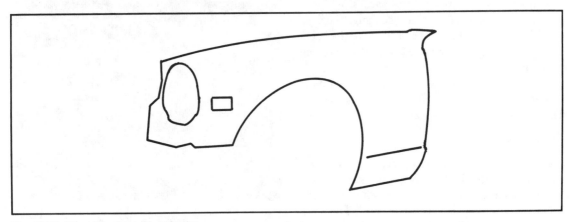

Fig. 13-24. Fiberglass front driver fender.

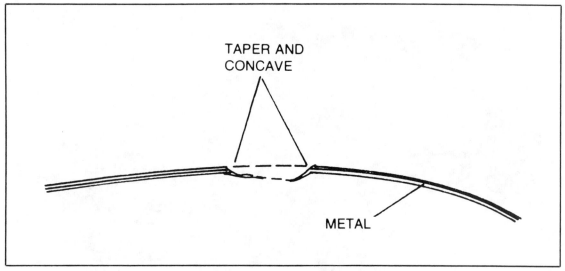

Fig. 13-25. Taper and concave the edge of the metal around the hole.

Finishing begins with removing excess filler material. Use a hand surfacing tool or remove the material by hand or power sanding. Progressively work down to finer sandpaper until the desired surface is achieved.

If any low spots or surface defects exist, mix and apply additional filler material. Allow the material to harden, then sand as necessary.

Minor imperfections such as pin-size holes and shallow scratches can be filled with glazing putty. The area is then ready for final sanding. Finally, prime and paint the area.

Customizing with Polyester and Epoxy Resins and Fiberglass Reinforcing Materials

Similar customizing work can also be done with polyester and epoxy resins and fiberglass reinforcing materials. When larger areas are to be filled in, this method is generally superior to using filler materials.

Proper preparation is extremely important if a good bond is to be achieved. Remove all paint and rust by sanding. If holes extend through the metal, concave the area around the holes to give a larger bonding surface. If a hole goes through the metal,

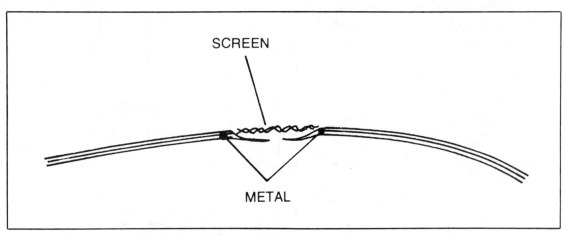

Fig. 13-26. Placement of screen in a depression in metal.

A & A FIBERGLASS, INC.

1534 NABELL AVENUE / ATLANTA, GEORGIA 30344
(404) 762-9631

A&A
URETHANE · PLASTIC
FIBERGLASS

Fig. 13-27. Car customized with molded fiberglass components (courtesy A & A Fiberglass, Inc.).

apply temporary or permanent backing as detailed earlier in this chapter for fiberglass body patches on metal.

Cut reinforcing material to desired sizes and shapes. Pour the required amount of resin into a

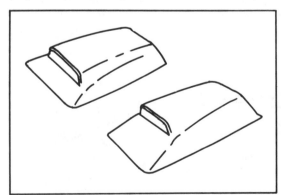

Fig. 13-28. Fiberglass air scoops.

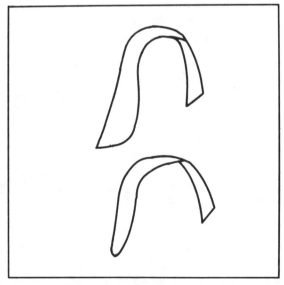

Fig. 13-29. Fiberglass fender flairs.

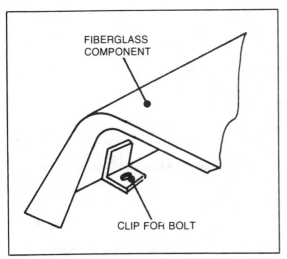

Fig. 13-30. Bolt-on mounting.

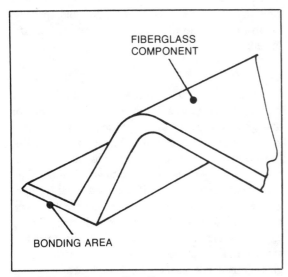

Fig. 13-31. Bond-on configuration.

mixing container and add catalyst or hardener according to manufacturer's directions for desired working time and working temperature.

Apply laminate as detailed for body repairs. Allow the completed laminate to cure.

Remove excess material by sanding or other means. Sand surface to the desired shape and contour. Progressively work down to finer sandpaper until the desired surface is achieved.

If any low spots or surface defects exist, mix and apply additional filler material. Allow the material to harden, then sand as necessary.

Minor imperfections such as pin-size holes and shallow scratches can be filled with glazing putty. The area is then ready for final sanding, priming, and painting.

Adding Molded Fiberglass Customizing Components

Molded fiberglass customizing components, such as air scoops and fender flairs, are now manufactured (see Appendix for suppliers) (Figs. 13-27 through 13-29). These are installed in a variety of ways such as with mechanical fasteners, epoxy gluing, and fiberglass bonding (Figs. 13-30 and 13-31). The fiberglassing methods detailed in this chapter for repairs and customizing can be used for installing these components.

Molded fiberglass custom tops are available for vans. Some of these are fixed tops. Others have canvas sides and can be lowered and raised. Typical installation includes cutting the top section out of a metal van top, putting the molded fiberglass top in place — usually with sealing compound or a gasket placed between metal and fiberglass — and attaching the top with mechanical fasteners. Custom components are also available for certain cars.

Chapter 14

Fiberglass Products in Kit Form

In addition to the fiberglassing repair kits discussed previously in this book, fiberglass products such as auto bodies, boats, travel trailers, and camper tops for vans are available in kit form. The basic parts of these kits are the molded fiberglass components. The factory does the molding operations, and you assemble the molds into a finished product.

A list of fiberglass product suppliers in kit form is included in the Appendix. Let's look at some examples of fiberglass products in kit form.

KIT CARS

Fiberglass kits are available for sports car, hot rod, dune buggy, and other types of car bodies. In most cases the bodies are designed to fit certain standard or modified automobile chassis.

Perfect Plastics Industries, Inc., Schreiber Industrial District, New Kensington, PA 15068, offers kits for the *Boss Bug* (Figs. 14-1 and 14-2) and the *Tuff Tub* (Figs. 14-3 and 14-4).

Boss Bug Kit

The Boss Bug kit fits a shortened standard VW sedan chassis. The kit comes with detailed step-by-step assembly and finishing instructions. The basic kit includes an extra heavy-duty fiberglass body (available in a number of colors, both glossy and metal-flake) with molded-in battery and spare tire compartments; fiberglass hood and leather-grained dash with wood-grained instrument insert; legal 7-inch all-chrome headlights, front parking lights, and black fender welt; polished aluminum windshield frame completely assembled with safety glass, windshield gaskets and all necessary hardware; two heavy-duty fiberglass bucket seats; chrome license plate light and brackets; and required plated nuts, bolts, washers, and spacers. Illustrated frame shortening instructions are included.

Tuff Tub Kit

The Tuff Tub kit fits a standard VW sedan chassis. No cutting, shortening, welding or control linkage modifications are necessary, making installation faster and easier than is the case with the Boss Bug kit described earlier. The kit includes a heavy-duty fiberglass body with molded-in color, black body undercoat, and molded-in black leather-grained inner liner; matching engine cover; silver metal-flake grill shell with black welt trim; 7-inch chrome headlight shells; polished aluminum windshield frame completely assembled with safety glass, windshield gaskets, and hardware; required plated nuts, bolts, and washers; two heavy-duty fiberglass bucket seat shells; chrome engine cover latches; fiberglass gas tank support saddle; steering column support bracket and extender; engine cover hinge; front shock tower adapters (required only for 1965 VW or earlier models); chrome automotive windlace for inside (passenger area) edges and engine cover trim; and detailed step-by-step assembly and finishing instructions. The shipping weight of the kit is about 240 pounds.

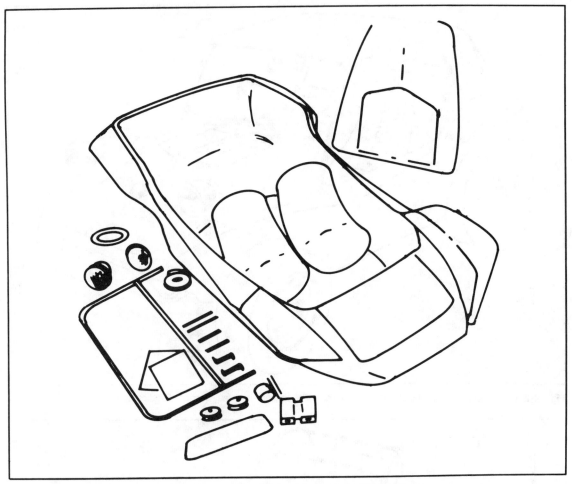

Fig. 14-1. Boss Bug fiberglass car body kit (courtesy Perfect Plastics Industries, Inc.).

Some features of both the Boss Bug kit and the Tuff Tub kit are:

☐ High quality heavy-duty moldings.
☐ Molded in color; no painting is required.
☐ Ease of assembly.
☐ Economically priced (write to manufacturer for current prices).
☐ No fiberglassing is required for assembly.

KIT BOATS

Several types and sizes of fiberglass boats are available in kit form. The basic idea is that you purchase factory-molded fiberglass hulls and other fiberglass and nonfiberglass components, and then all or part of the assembly yourself. This is generally much easier to do, and the resulting boat is likely to be far superior to a fiberglass boat built from scratch. This is probably an understatement. Building a fiberglass boat from scratch takes considerable talent and commitment if a successful boat is to result; in comparison, assembly of a fiberglass kit boat can be easy.

There are many concepts, options, and possibilities available. At one extreme there is the very complete kit. You get all the parts, which are pre-shaped to fit, fasteners, bonding and sealing chemicals and materials, hardware, and so on. You also get complete assembly instructions. In short, you

Fig. 14-2. Boss Bug car body kit installed on shortened VW chassis (courtesy Perfect Plastics Industries, Inc.).

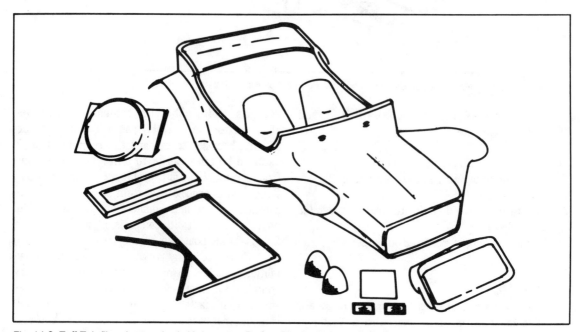

Fig. 14-3. Tuff Tub fiberglass car body kit (courtesy Perfect Plastics Industries, Inc.).

Fig. 14-4. Tuff Tub car body kit installed on standard VW sedan chassis (courtesy Perfect Plastics Industries, Inc.).

get everything you need, minus tools, for a complete boat. You assemble the parts.

At the other extreme you purchase a bare fiberglass hull. To finish this out, you purchase your own materials and supplies. Obviously, this requires more time, work, dedication, talent, and experience. It can also mean the most boat for the least money. It's essentially boatbuilding starting with a bare hull. The advantage over building from scratch is that you start with a (hopefully) sound and proven hull, something that already has the basic shape of a boat. Building from scratch means that you start with a set of plans and work from them.

Between the two extremes—the complete kit of all necessary parts and the bare hull—are many options. With the bare hull, you can purchase some or all of the remaining molded fiberglass components such as the deck and cabin structure, hatches, rudder, hull liner, cabin liner, and so on. These can be purchased with or without part of the assembly done for you at the factory, such as the hull and deck joined and bonded together. You can purchase various other component kits such as preshaped interior wood, ballast, exterior trim and hardware, engine, mast and rigging, and so on. These can be purchased with or without part of the assembly and/or installation being done at the factory. For example, you might want to have the ballast installed at the factory. You might want the engine installed, and so on. Some manufacturers offer some or all of these options.

Fig. 14-5. The Leeward 16 fiberglass sailboat is offered in kit form by Luger Industries, Inc. (courtesy Luger Industries, Inc.).

Fig. 14-6. The Voyager 30 is a 30-foot trailerable sailboat that is available in fit form (courtesy Luger Industries, Inc.).

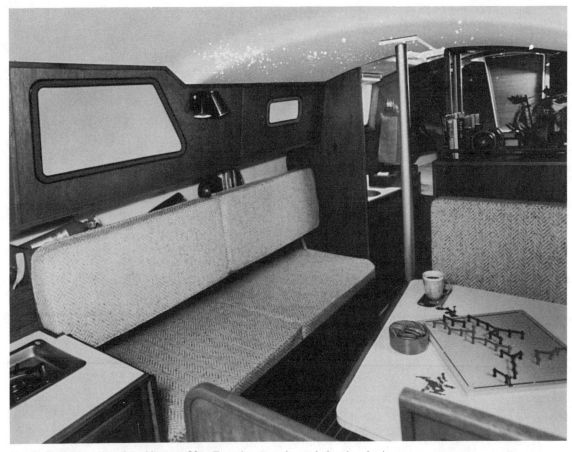

Fig. 14-7. Interior view of the Voyager 30 sailboat (courtesy Luger Industries, Inc.).

Another popular option is what is called a power-away or sail-away boat. These boats are generally complete except for the interiors, which are bare. The boats can usually be launched and used; the interiors can be completed as time and finances permit.

Some manufacturers offer their boats in a series of kits, which you don't have to purchase all at once. You can buy them as you can afford them, thus avoiding finance charges.

Kit Boats from Luger Industries, Inc.

Luger Industries, Inc., 3800 West Highway 13, Burnsville, MN 55337, has been offering kit boats for 28 years. Their kits are very complete, well thought out, and easy to assemble. By com-

pleting one of their kit boats, it's possible to save about half the cost of a comparable factory finished boat. They offer a complete line of trailerable power and sail boats from 16 to 30 feet in length.

The smallest boat in their line is the *Leeward 16* (16 feet long) sailboat (Fig. 14-5). Completion time for this boat is approximately 10 to 15 hours.

The Voyager 30, a 30-foot trailerable sailboat, is the largest sailboat in their line (Fig. 14-6). Figure 14-7 shows a view of the interior. The basic assembly concept is shown in Fig. 14-8. The deck plan is shown in Fig. 14-9; the interior plan is Fig. 14-10; and the profile is Fig. 14-11. Figure 14-12 shows the boat on a trailer. The trailers are also available from Luger Industries, Inc.

Their powerboats range in size from 21 to 30 feet. Figure 14-13 shows the Vagabond, which is

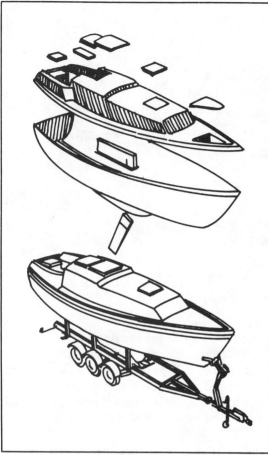

Fig. 14-8. Assembly concept for the Voyager 30 sailboat (courtesy Luger Industries, Inc.).

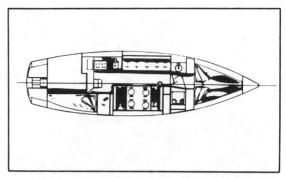

Fig. 14-10. Interior plan view of the Voyager 30 sailboat (courtesy Luger Industries, Inc.).

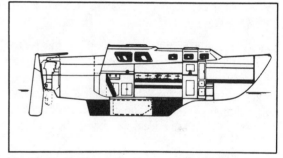

Fig. 14-11. Profile of the Voyager 30 sailboat (courtesy Luger Industries, Inc.).

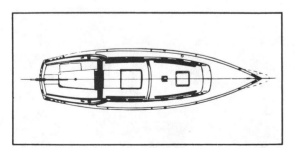

Fig. 14-9. Deck plan of the Voyager 30 sailboat (courtesy Luger Industries, Inc.).

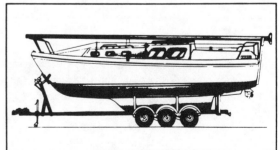

Fig. 14-12. The Voyager 30 is designed for trailering (courtesy Luger Industries, Inc.).

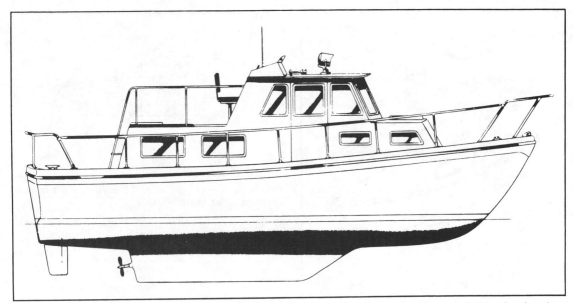

Fig. 14-13. The Vagabond is a trailerable 30-foot powerboat that is available in kit form from Luger Industries, Inc. (courtesy Luger Industries, Inc.).

30 feet long. It has a displacement hull that is easily driven at a cruising speed of about 7 knots by a 40-horsepower diesel engine. With the designed 60-gallon fuel tankage, the boat has a cruising range of up to 500 miles. The interior accommodations (Fig. 14-14) are ideal for living aboard.

Dreadnought 32

A popular world cruising type sailboat that is available as a bare hull, with or without components for completing it, and at any desired stage of

completion is the *Dreadnought 32*. It's offered by Dreadnought Boatworks, P.O. Box 221, Carpinteria, CA 93013. Figure 14-15 shows a basic hull and deck and cabin version with the deck joined to the hull. Figure 14-16 shows a molded fiberglass rudder being assembled to the hull.

Building from a Bare Hull

A more ambitious undertaking is to build from a bare hull. Figure 14-17 shows the installation of a bulkhead, which is bonded in place with strips of

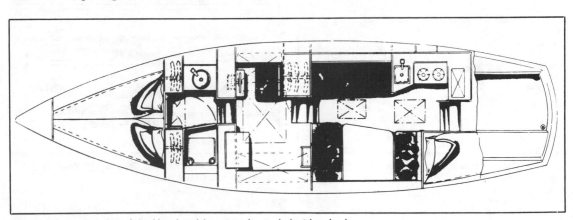

Fig. 14-14. Interior plan of the Vagabond (courtesy Luger Industries, Inc.).

Fig. 14-15. Hull and deck moldings of the popular Dreadnought 32.

Fig. 14-16. Molded fiberglass rudder being assembled to hull of Dreadnought 32.

Fig. 14-17. Bulkhead installed in a Dreadnought 32 hull.

fiberglass as detailed previously in this book. In Fig. 14-18, a wood shelf clamp is being attached to the fiberglass hull for mounting a wood deck in place. Figure 14-19 shows the wood shelf clamp after it has been attached to the fiberglass hull. Figure 14-20 shows deck construction attached to a shelf clamp. The cabin was later added to it (Fig. 14-21). The cabin construction consisted of mahogany beams, plywood planking, and a fiberglass covering. This type of construction is considerably more difficult than using the factory molded deck and cabin components.

A list of kit boat suppliers is given in the Appendix. While most of these kits do not require

Fig. 14-18. Attaching wood shelf clamp for mounting a wood deck to a fiberglass hull.

Fig. 14-20. Deck construction attached to shelf clamp.

Fig. 14-19. The wood shelf clamp has now been attached to the fiberglass hull.

Fig. 14-21. Plywood cabin is covered with fiberglass.

any additional fiberglass molding (this is done by the manufacturer), most do require fiberglass bonding and other fiberglassing skills. These are essentially the same as the fiberglassing repair skills and techniques detailed throughout this book.

Space allows only a brief introduction to fiberglass kit boats here. A suggested source of additional information is my book, *Boatbuilding form Fiberglass Hulls and Kits,* available from Solipaz Publishing Company.

OTHER FIBERGLASS KITS

Other fiberglass kits, including travel trailers, van tops for converting to standup headroom campers, and other products, are also available. Most of these can be assembled without fiberglassing. A list of suppliers is given in the Appendix.

Chapter 15

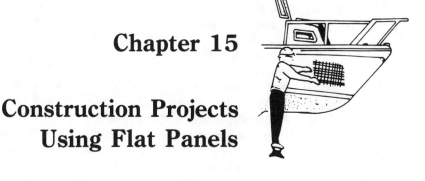

Construction Projects Using Flat Panels

A variety of interesting projects are possible using flat panels of fiberglass. The basic idea is to form flat panels of fiberglass, which in turn can be used as panels for room screens, skylights, cabinet doors, and other similar constructions, or joined together with fiberglass bonding strips to form boxes, tanks, and other projects.

FORMING FLAT PANELS OF FIBERGLASS

All of the projects detailed in this chapter require flat panels of fiberglass. These can be purchased preformed, or you can mold your own with one or both surfaces against the mold surface. When one smooth surface is required, use contact molding. When a molded surface is required on both sides of the panel, use pressure molding.

Single Flat Molds

Flat molds can be formed from a variety of materials. Plywood is frequently used for making the mold surface. The first step is to cut the plywood to the desired size. Generally this is slightly larger than the desired panel size. If a variety of sizes are to be molded, make the plywood larger than the largest desired size.

Prepare the plywood surface by applying a thin coat of polyester resin and then sanding and polishing this to a smooth surface. Lacquer can be used in a similar manner.

A better method, I have found, is to cement a plastic laminate to the plywood (Fig. 15-1). This

can be attached with contact cement. Use the kind formulated for attaching plastic laminates to wood. Apply a thin layer to both plywood surface and the bonding side of the plastic laminate and allow the cement to dry. Position a piece of construction paper over the dry contact cement on the plywood. Then position the plastic laminate cement-side-down on the paper. When you have the plastic laminate lined up like you want it, slide the paper out. Use a block of wood and a hammer to lightly tap the laminate in place. Wax and mold release can then be applied to the plastic laminate to prepare it for molding.

The mold can be placed directly over protective paper on a floor. It often is more convenient, however, to work with the mold placed on a table or other suitable platform.

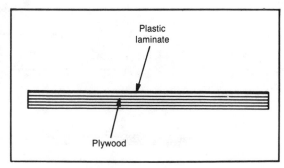

Fig. 15-1. Cement plastic laminate to plywood to use it as a flat mold.

Contact Molding Flat Panels

To mold a flat panel, you need a flat mold, as detailed above, a paste-type mold release, polyester gel coat resin, laminating and/or finishing or general-purpose polyester resin, catalyst for the resin, mixing cups or containers, mixing sticks, a small brush for applying the resin, and fiberglass reinforcing material (type, weight, and number of layers depends on the desired laminate and the specific project). For larger size panels, squeegees and laminating rollers are also useful. You will also need heavy paper to protect the work area, which can be on a bench, floor, or concrete surface outside. Wear proper protective clothing and safety equipment (Chapter 3) and follow all safety rules (Chapter 4).

Place the flat mold on the protective paper with the mold surface upward. Apply a coat of the paste-type mold release, using a clean cloth to spread a thin layer of the release over the entire mold surfaces. The purpose of this is to keep the molding from sticking so that it can be removed after it has cured.

The next step is to apply the gel coat on the mold surfaces over the mold release. Open the container of gel coat polyester resin. This resin can be clear or any desired color, or you can add color pigments formulated for polyester gel coat resin to it. Carefully stir the resin with a clean mixing stick. When removing the stick, allow as much of the resin to drain off the stick as possible. Clean off the remaining resin from the stick.

Pour the required amount (this depends on area to be covered; usually about 1 ounce per square foot is required) of the gel coat resin into a mixing cup. Replace the lid on the gel coat resin container and set it aside.

Next, add the catalyst to the resin. This is approximately 2 percent by volume, which is usually used for polyester gel coat resin at 75 degrees Fahrenheit. This can vary for the particular brand of gel coat resin being used, however. Follow the manufacturer's directions. Wait about 30 seconds, then brush a smooth thin layer of the catalyzed gel coat resin onto each of the mold surfaces. Use a continuous motion with the brush (Fig. 15-2) rather than painting back and forth. The gel coat should have a thickness of only .02 to .03 inch. A thicker layer will be brittle and subject to cracking and crazing. When you have finished applying the

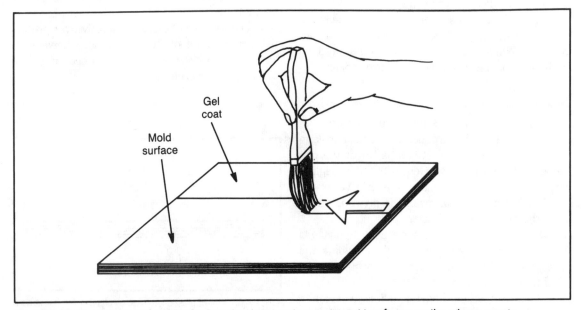

Fig. 15-2. Brush a thin layer of catalized polyester gel coat resin onto the mold surface over the release agent.

coat of resin to the plywood pieces and everything looks okay, clean the brush with acetone. Allow the resin to harden. Gel coat resin usually doesn't contain a wax additive so that the laminate can be laid up directly over this with no additional surface preparation.

The next step is to laminate the layer of fiberglass reinforcing material (which is usually mat) to the gel coat. Beginners are advised to apply one layer of reinforcing material at a time and allow this to set before applying the next layer. As experience is gained, two or more layers (up to a maximum of about 6 ounces of reinforcing material per square foot) can be applied wet. This is usually more difficult than applying a single layer at a time but makes the work faster.

In any case, the next step is to calculate the amount of resin required for saturating the reinforcing material that is to be applied wet at this time. Methods for determining the amount of resin required for various weights of reinforcing material were given in Chapter 7.

Pour the required amount of resin into a mixing cup. You will probably need about 7½ to 10 minutes of working time to apply the resin.

Methods for determining the amount of catalyst to add to give desired working times at various working temperatures were given in Chapter 7.

Add the required amount of catalyst to the resin and stir. Wait about 30 seconds. Then, using the brush, apply a coat of resin over the gel coat layer that was applied to the mold surface and allowed to set previously.

Position the first layer of fiberglass reinforcing material (usually a mat layer) over the wet resin (Fig. 15-3). Center it, press it down, and smooth it out. For mat, use a dabbing action rather than a brushing action to avoid lumping up the mat (Fig. 15-4).

Saturate the reinforcing material with resin (Fig. 15-5). Spread the resin to saturate the area as evenly as possible, including areas of the reinforcing material that extend slightly beyond the desired size of the finished panel. (These will be trimmed off later after the laminate has cured.)

For larger panels, use a squeegee (Fig. 15-6) or laminating roller (Fig. 15-7) to remove air bubbles. Small and medium size panels can be made without joints in the individual layers of reinforcing material by purchasing the reinforcing material in

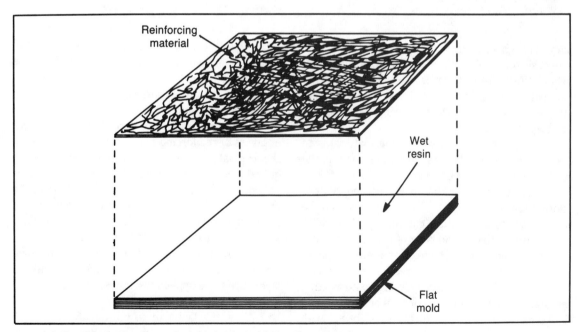

Fig. 15-3. Position the first layer of reinforcing material over the wet resin.

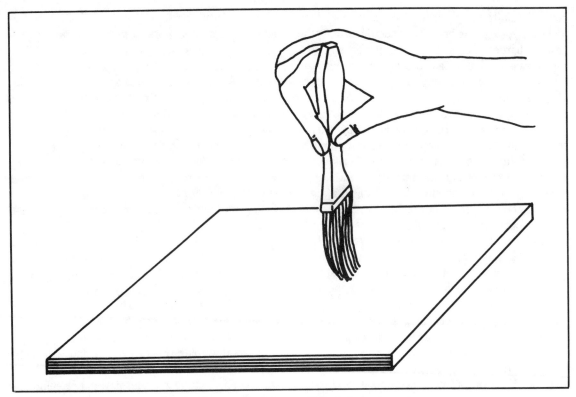

Fig. 15-4. Use a dabbing action with the brush to wet out the fiberglass mat.

large enough sizes. If joints are required, however, the basic technique is to overlap the pieces about 2 inches (Fig. 15-8). If mat is used, work down the laminate to a nearly even thickness with a dabbing action of the brush when applying resin to the overlap (Fig. 15-9).

If a second layer of reinforcing material is to be applied wet over the first layer, position it in the wet resin and press it down and smooth it out. The brush can be used for this. If the second layer is cloth rather than mat, use a brushing action but be careful not to unravel the cut edges of the cloth.

Saturate the second layer of reinforcing material with resin. Spread the resin as evenly as possible. Saturate all areas of the reinforcing material.

Apply all layers of reinforcing material that are to be applied wet at this time. When everything looks okay, clean the resin from the brush with acetone.

It works best if laminating resin is used for all but the final layer of resin on the backside of the laminate. The final layer should be sanding or finishing resin to give a surface that is not sticky so that it can be sanded.

Allow the laminate to cure. If additional layer or layers of reinforcing material are to be added to the laminate, measure out and catalyze the required amount of polyester resin. Brush on a layer of wet resin over the cured laminate. Position the first additional layer of fiberglass reinforcing material over the wet resin to center it, press it down and smooth it out using the brush.

Saturate the reinforcing material with resin. Spread the resin as evenly as possible. If additional layers are to be applied wet, these at this time in a similar manner.

Allow the laminate to thoroughly cure, then remove it from the mold. Even though the mold

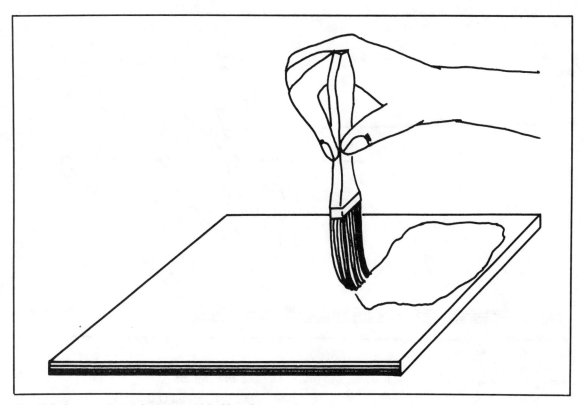

Fig. 15-5. Saturate the reinforcing material with resin.

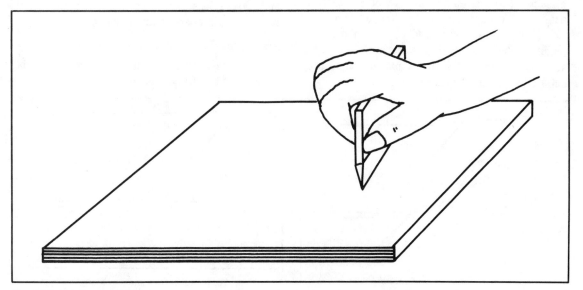

Fig. 15-6. Using a squeegee.

215

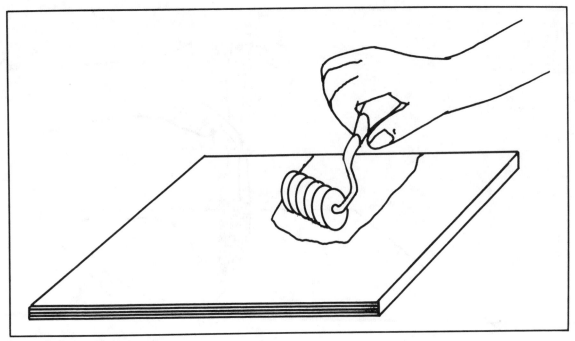

Fig. 15-7. Using a laminating roller.

release was used, it still might be necessary to pry the laminate loose with a putty knife or other similar tool.

This completes the chemical part of the construction. Mark the pattern for cutting the fiberglass panel to desired size and saw off excess laminate. Sawing methods are detailed in Chapter 6.

Next, file and sand the edges smooth (this might not be necessary for some projects). Polish and buff the gel coat surfaces and (for some projects) edges of the laminate.

Double Flat Molds

The bottom half of the mold is the same as for the single flat mold. The matching upper half of the mold is constructed in the same manner, usually by attaching a plastic laminate to the plywood. Wax and mold release are then applied to the plastic laminate to prepare it for molding.

Pressure Molding Flat Panels

To pressure mold a flat panel in the double mold (Fig. 15-10), you need a double flat mold, as

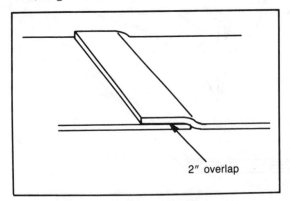

Fig. 15-8. Reinforcing material joined with overlap.

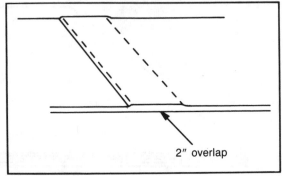

Fig. 15-9. Work down the mat overlap to near even thickness.

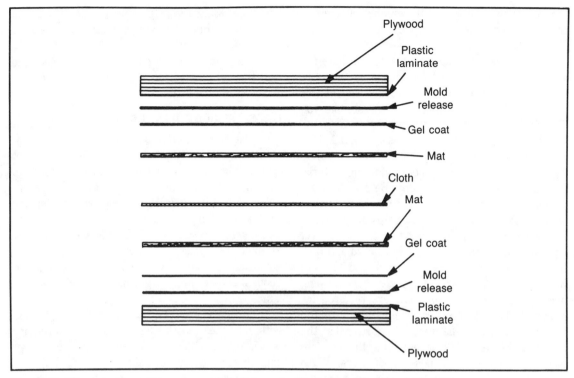

Fig. 15-10. Use of double mold to pressure mold flat panel with smooth surfaces on both sides.

detailed, cement building blocks or other suitable weights for applying pressure to the upper mold surface, a paste-type mold release, polyester gel coat resin, laminating and/or finishing or general-purpose polyester resin, catalyst for the resin, mixing cups, mixing sticks, a small brush for applying the resin, and fiberglass reinforcing material (type, weight, and number of layers depends on the desired laminate and the specific project). For larger size panels, squeegees and laminating rollers are also useful. You also need heavy paper to protect the work area. Wear proper protective clothing and safety equipment (Chapter 3) and follow all safety rules (Chapter 4).

Place the two sections of the flat mold on the protective paper with the mold surfaces upward. Apply a coat of the paste-type mold release, using a clean cloth to spread a thin layer of the release over the entire mold surface of each mold half. The purpose of this is to keep the molding from sticking so that it can be removed after it has cured.

The next step is to apply the gel coat on the mold surfaces over the mold release on both parts of the mold. This is done in the same manner as detailed for a single flat mold, except that the gel coat is applied separately to both parts of the mold. Gel coat resin usually doesn't contain a wax additive so that the laminate can be laid up directly over this with no additional surface preparation.

The next step is to laminate the first layer of fiberglass reinforcing material (which is usually mat) to the gel coat of the bottom half of the double mold. In some cases, only one layer will be required. Other projects require panels of two or more layers. The double mold method detailed here requires applying all layers wet. As a rule, the maximum practical thickness for this method of molding is about six layers of reinforcing material. This quantity of reinforcing material can be applied wet by an experienced fiberglasser. Beginners should not attempt more than two or three layers until experience is gained. The thickness and types of reinforcing material and the number of layers in the laminate depend on the specific project you are making, as detailed later in this chapter.

In any case, the next step is to calculate the amount of resin required for saturating the reinforcing material that is to be applied wet at this time (see Chapter 7). Catalyze the resin like you did for the single flat mold.

Lay up the laminate, again as was detailed for the single flat mold, except that all layers must be laid up wet. Apply all layers of reinforcing material that are to be used at this time. Before the laminate has had a chance to set up, turn the upper half of the mold so that the cured gel coat surface is face down and press it in place on top of the wet layer of cloth. Apply pressure downward and rock the upper section of the mold back and forth to work out any air. Then place the cement building block or blocks (more than one are required for larger mold areas) or other suitable weight on top of the upper half of the mold. Make sure that the upper half of the mold is level so that the laminate will have an equal thickness in all areas. When everything looks okay, clean the resin from the brush with acetone.

Allow the laminate to thoroughly cure, then remove it from the mold. Even though the mold release was used, it still might be necessary to pry the laminate loose with a putty knife or other similar tool.

This completes the chemical part of the construction. Mark the pattern for the desired size of panel on the fiberglass molding and saw off excess laminate.

File and sand the edges smooth. Polish and buff the gel coat surfaces and (for some projects) edges of the laminate.

USES OF FLAT PANELS

Flat panels can be used for a variety of constructions. A serving tray (Fig. 15-11), cabinet doors (Fig. 15-12), sliding cabinet doors (Fig. 15-13), folding screens (Fig. 15-14), and room dividers (Fig. 15-15) are interesting projects that use decorative fiberglass panels. A fiberglass panel of two layers of 1½-ounce fiberglass with a layer of 10-ounce fiberglass cloth sandwiches in between or equivalent thickness will suffice for areas up to about 2 square feet. Larger panels require thicker laminates.

For some projects, the panels can be molded with only one smooth side. Other projects require panels that are double molded so that they have smooth surfaces on both sides.

The appearance of the panels can often be improved by laying down colored burlap or cotton cloth over a wet layer of resin applied to the gel coat in the mold before the first layer of fiberglass reinforcing material is added. The material should first be washed with a detergent and pressed with a hot iron to remove all wrinkles. It is then ready to place face down in the wet resin.

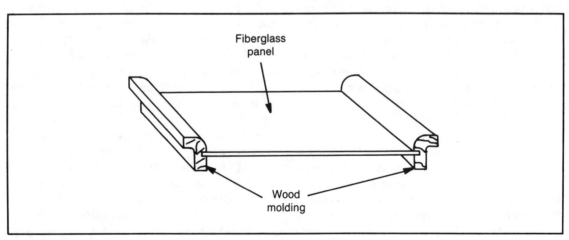

Fig. 15-11. Serving tray.

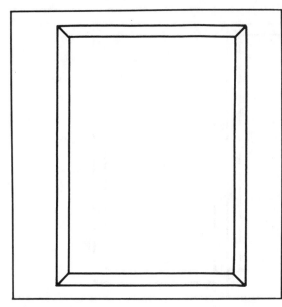

Fig. 15-12. Cabinet door.

Frames for fiberglass panels can be made from wood (Fig. 15-16) or aluminum (Fig. 15-17) moldings.

LAMINATING A FLAT
FIBERGLASS PANEL DIRECTLY TO WOOD

Fiberglass laminated to plywood makes an ideal decorative and protective surface, such as for a tabletop for outdoor furniture (Fig. 15-18). Unlike the other projects using flat fiberglass panels, the laminate is bonded directly to the plywood base (Fig. 15-19).

The plywood base that the fiberglass is to be bonded to is used for the bottom half of the mold in the same way as for pressure molding flat panels. This time, however, no release or gel coat will be applied so that the fiberglass is permanently bonded to the plywood base. The upper half of the mold is prepared the same as for regular pressure molding. Wax and mold release are applied to the plastic laminate to prepare it for molding.

To pressure mold a layer of fiberglass mat to a plywood base, you will need the upper half of a pressure mold. Cement building blocks or other suitable weights will be used to apply pressure to

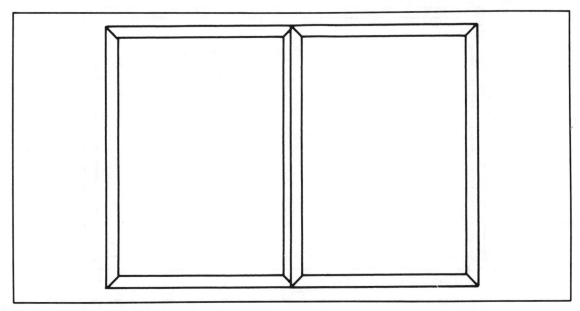

Fig. 15-13. Sliding cabinet doors.

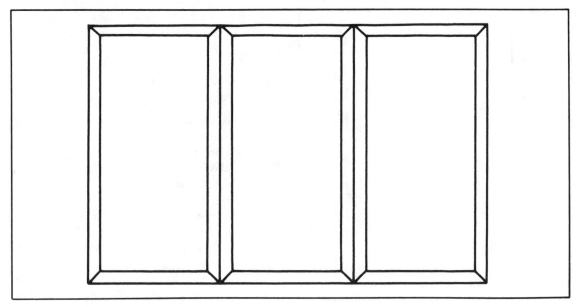

Fig. 15-14. Folding screen.

the upper mold surface, or C-clamps can be used around the edges to clamp the two pieces of wood together under pressure. You also need a paste-type mold release, polyester gel coat resin, laminating and/or finishing or general-purpose polyester resin, catalyst for the resin, mixing cups (the kind sold at drug stores with ounce markings on the sides are recommended), short and long mixing sticks, protective paper, a small brush for applying the resin, and 1½ ounce fiberglass mat reinforcing

Fig. 15-15. Room divider.

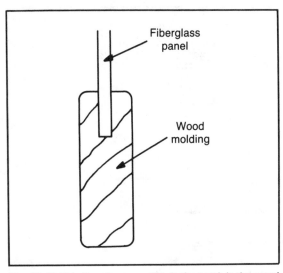

Fig. 15-16. The fiberglass panel fits in the notch in the wood molding.

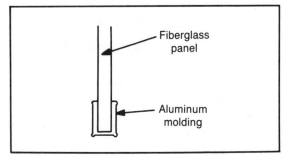

Fig. 15-17. The fiberglass panel fits in the notch in the aluminum molding.

Place the upper section of the flat mold on the protective paper with the mold surfaces upward. Apply a coat of the paste-type mold release, using a clean cloth to spread a thin layer of the release over the entire mold surface. This will keep the molding from sticking so that it can be removed

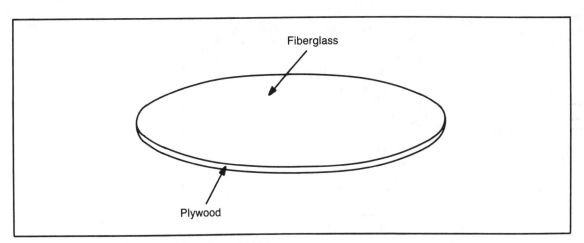

Fig. 15-18. Fiberglass pressure molded to plywood tabletop.

material large enough to cover the tabletop with at least a ½-inch overlap all the way around the plywood base. For large tabletops, separate pieces of fiberglass mat can be used with 2-inch overlaps at joints, although if at all possible, use a single large piece. For larger tabletops, squeegees and laminating rollers are also useful. Wear proper protective clothing and safety equipment (Chapter 3) and follow all safety rules (Chapter 4).

after it has cured. Next, apply the gel coat on the mold surface you did over the mold release as for single flat molds.

The next step is to laminate the fiberglass mat to the plywood base. Calculate the amount of resin required for bonding the mat to the plywood and saturating it with resin.

Add the required amount of catalyst to the resin and stir. Wait about 30 seconds. Then, using

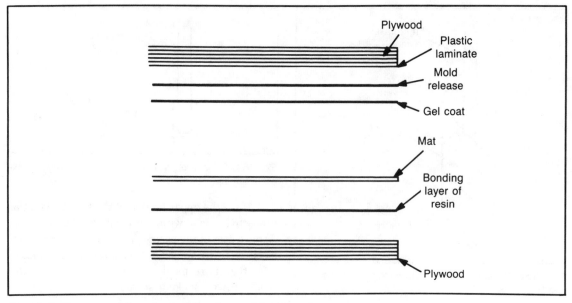

Fig. 15-19. Pressure molding a layer of fiberglass mat to plywood.

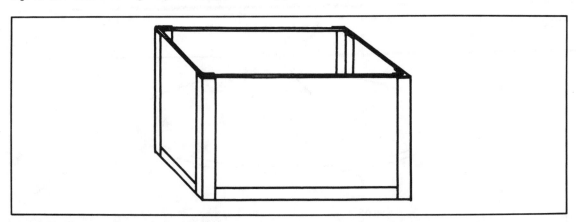

Fig. 15-20. Fiberglass box made from flat panels.

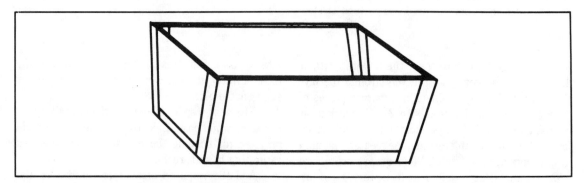

Fig. 15-21. Planter box made from flat fiberglass panels.

222

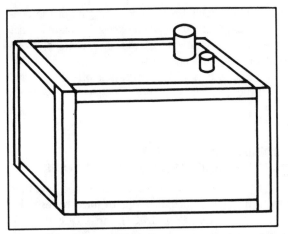

Fig. 15-22. Fiberglass tank made from flat panels.

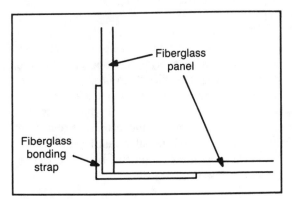

Fig. 15-23. Flat fiberglass panels joined with bonding strap on one side.

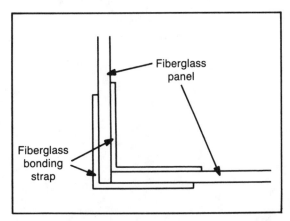

Fig. 15-24. Fiberglass panels joined with fiberglass bonding straps on both sides.

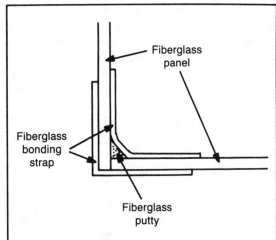

Fig. 15-25. Use fiberglass putty fill to give a rounded inside corner.

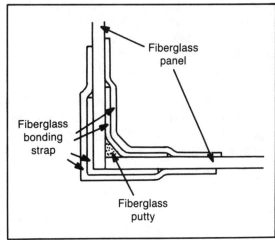

Fig. 15-26. Arrangement of multiple layers of bonding straps.

the brush, apply a bonding coat of resin to the plywood base. Position the fiberglass mat reinforcing material over the wet resin (see Fig. 15-3) until it is centered; press it down and smooth it out.

Saturate the reinforcing material with resin (see Fig. 15-5). Spread the resin as evenly as possible. Saturate all areas evenly, including areas of the reinforcing material that extend slightly beyond the tabletop plywood. These will be trimmed off later after the laminate has cured.

223

For larger tabletops, use a squeegee (see Fig. 15-6) or laminating roller (see Fig. 15-7) to remove air bubbles. The tabletop is best covered with a single piece of mat reinforcing material. If joints are required, however, the basic technique is to overlap the pieces about 2 inches (see Fig. 15-8). The mat laminate can be worked down to a nearly even thickness with a dabbing action of the brush when applying resin to the overlap (see Fig. 15-9).

Before the laminate has had a chance to set up, turn the upper half of the mold so that the cured gel coat surface is face down, and press it in place on top of the wet layer of mat. Apply pressure downward and rock the upper section of the mold back and forth to work out any air. Then place the cement building block or blocks (more than one are required for larger mold areas) or other suitable weight on top of the upper half of the mold. Make sure that the upper half of the mold is level so that the laminate will have an equal thickness in all areas. An alternate method is to use C-clamps around the edge of the tabletop. When everything looks okay, clean the resin from the brush with acetone.

Allow the laminate to thoroughly cure. Remove the upper mold half from it. Even though the mold release was used, it still might be necessary to pry the mold loose.

This completes the chemical part of the construction. Mark and saw off excess laminate so that the fiberglass is even with the edges of the plywood all the way around.

File and sand the edges smooth. Polish and buff the gel coat surface of the fiberglass laminate.

JOINING SEPARATE PANELS WITH FIBERGLASS BONDING STRAPS

Flat fiberglass panels can be joined together with fiberglass bonding strips to form boxes (Fig. 15-20), planters (Fig. 15-21), tanks (Fig. 15-22), and a variety of other projects. The fiberglass bonding straps can be on one (Fig. 15-23) or both (Fig. 15-24) sides, depending on the particular construction.

After the panels have been cut to size, the individual panels are then clamped or otherwise held in position for applying the fiberglass bonding strips. Because it is difficult to fiberglass in sharp corners, fiberglass putty is often applied to give a rounded surface (Fig. 15-25). If more than one layer of fiberglass reinforcing material is to be used for a bonding strap, the narrowest layer is usually applied first, with each additional layer overlapping the one below (Fig. 15-26).

Chapter 16

Fiberglass Laminating Over Rigid Foam

A variety of interesting projects are possible by fiberglass laminating over rigid foam forms. The basic idea is to use rigid foam plastic, which is easily shaped, as a form for laying up a fiberglass laminate. In most cases, the rigid foam becomes part of the finished construction, and the fiberglass laminate is applied to all sides of the foam. Constructions can include boxes, planters, and even boat hulls. This method is ideal for constructs where only one is required. The main disadvantage to this method is that the finished side is not molded, making extensive fairing and sanding necessary if a smooth surface is desired. For many uses, however, this is still the most practical method when one considers the alternative of making a contact mold (as detailed in Chapter 17).

RIGID FOAM FORMS

All of the projects detailed in this chapter require rigid foam forms. The foam used must be compatible with the resin that will be used for the fiberglass lay-up.

Rigid Foamed Plastics

Rigid foamed plastics are detailed in Chapter 2. The three basic types used as forms for fiberglass lay-ups are polystyrene, polyurethane, and polyvinyl chloride (PVC). All three types can be used with epoxy resin. Only polyurethane and polyvinyl chloride are compatible with polyester resins. Rigid foamed plastics are available in various densities in sheet and block forms.

Making Rigid Plastic Forms

The basic idea is to shape the rigid foam to the desired shape of the finished construction, minus the thickness of the fiberglass skins to be applied. The form also becomes the core, for example, when a surfboard is shaped from a block of foam (Fig. 16-1). It is shaped to the desired finished size, then the fiberglass laminate is applied.

A box (Fig. 16-2) makes an ideal starting project. Cut the bottom and four sides from a ⅜-inch thick sheet of rigid plastic foam to the dimensions shown in Figure 16-3. Epoxy glue the parts to form the box shape (Fig. 16-4). Because fiberglass does not form well to sharp inside or outside angles, use a fiberglass putty fill on inside corners to give a rounded contour (Fig. 16-5) and round outside angles with a file or coarse sandpaper (Fig. 16-6). The edges of the lip of the box should also be rounded (Fig. 16-7).

APPLYING FIBERGLASS LAMINATE TO RIGID FOAM FORM

To apply a fiberglass laminate to the box form, you will need a rigid foam form, laminating and finishing or general-purpose polyester resin, catalyst for the resin, mixing cups and sticks, a small brush for applying the resin, and squeegees and laminating rollers for larger areas. The type, weight, and number of layers of fiberglass reinforcing material depends on the desired laminate and the specific project. For the box project, one layer

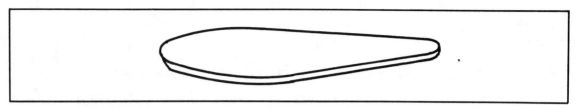

Fig. 16-1. Core form for surfboard shaped from rigid plastic foam.

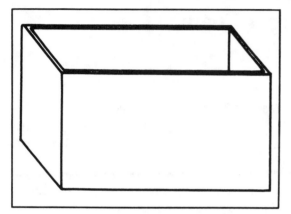

Fig. 16-2. Fiberglass box layed up over rigid foam plastic core.

of 1½ ounce fiberglass mat to cover the inside and outside surfaces with extra for corner overlaps is required.

Protect the work area with heavy paper. Wear proper protective clothing and safety equipment and follow all safety rules.

Place the box form over protective paper on the floor, a table, or other suitable platform. Face the surface where you will be applying the laminate upward. For the box, do the inside bottom surface first, followed by the inside sides, the lip, the outside sides, and finally the bottom. Allow the bottom inside surface to set up before doing one of the inside sides, and so on. This method allows you to fiberglass on a flat surface, with the form turned in

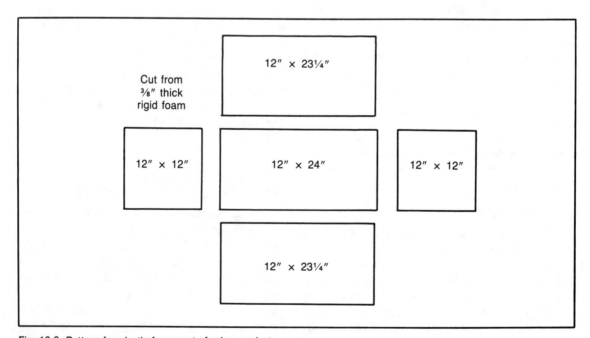

Cut from ⅜" thick rigid foam

12" × 23¼"

12" × 12"

12" × 24"

12" × 12"

12" × 23¼"

Fig. 16-3. Pattern for plastic foam parts for box project.

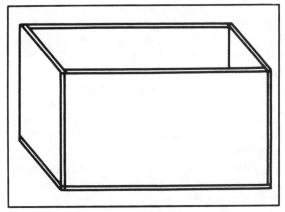

Fig. 16-4. Epoxy glue foam core sections together.

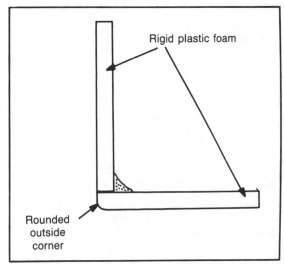

Fig. 16-6. Round outside corner.

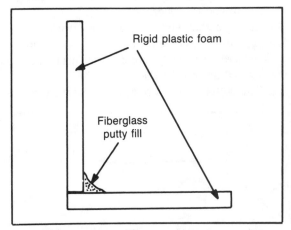

Fig. 16-5. Use fiberglass putty fill on inside corners to give a rounded contour.

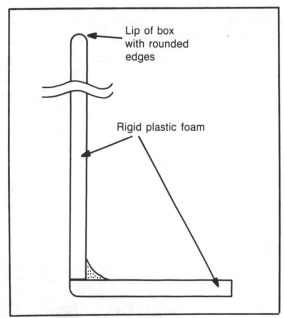

Fig. 16-7. Round edges of lip of box.

a new direction for each lay-up. Because the fiberglass is to be bonded to the foam core, no mold release is applied.

To laminate the first section of fiberglass skin in place, cut the fiberglass mat to size. For example, for the inside bottom of the box, this should be the size of the inside bottom plus an inch extra all the way around to extend upward on the sides (Fig. 16-8). Dry fit the mat first.

The next step is to laminate the layer of fiberglass reinforcing material rigid foam plastic. Beginners are advised to apply one layer of reinforcing material at a time and allow this to set before applying the next layer. This isn't a problem with

the box project as only one layer of mat is used. Many projects, however, require two or more layers of reinforcing material. As experience is gained, two or more layers (up to a maximum of about 6 ounces of reinforcing material per square foot) can be applied wet. This is usually more diffi-

227

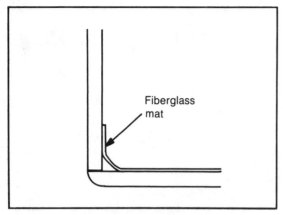

Fig. 16-8. Inside bottom section of fiberglass mat extends upward on sides about an inch all the way around.

cult than applying a single layer at a time but makes the work faster.

In any case, next calculate the amount of resin required for bonding and saturating the reinforcing material that is to be applied wet at this time (see Chapter 7). Pour the required amount of resin into a mixing cup. You will probably need about 7½ to 10 minutes of working time to apply the resin. Determine the amount of catalyst to add to give the desired working time at your working temperature.

Add the required amount of catalyst to the resin and stir. Wait about 30 seconds. Then, using the brush, apply a coat of resin over the rigid foam plastic in the area where the first section of fiberglass mat is to be applied.

Position the first layer of fiberglass reinforcing material over the wet resin until it is centered; press it down and smooth it out. For mat, use a dabbing action rather than a brushing action to avoid lumping up the mat (Fig. 16-9). Work the mat in place so that the edges extend upward about an inch around the sides.

Saturate the reinforcing material with resin (Fig. 16-10). Spread the resin as evenly as possible until all areas are evenly saturated, including areas that extend upward on the sides.

The bottom of the box is easily covered with a

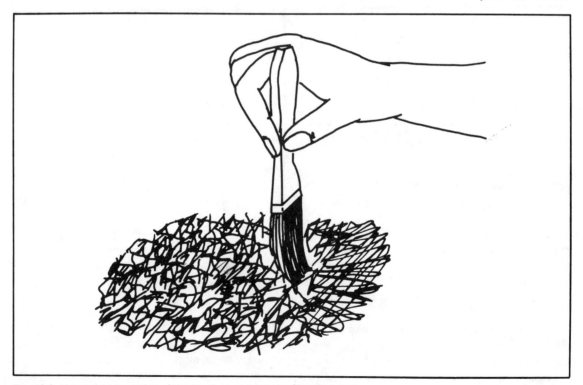

Fig. 16-9. Use a dabbing action of the brush to wet out the fiberglass mat.

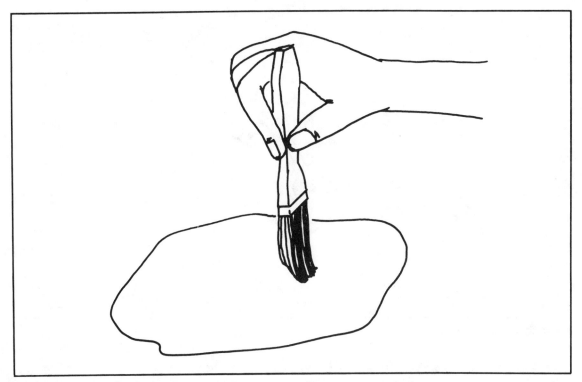

Fig. 16-10. Saturate the reinforcing material with resin.

single piece of fiberglass mat. Apply the resin with a brush. For larger panels, use a squeegee (Fig. 16-11) or laminating roller (Fig. 16-12) to remove air bubbles. Small and medium size areas can be covered without joints in the individual layers of reinforcing material by purchasing the reinforcing

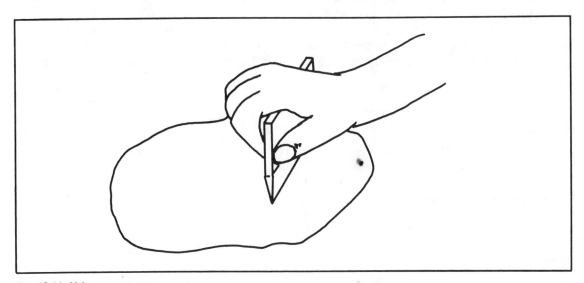

Fig. 16-11. Using a squeegee.

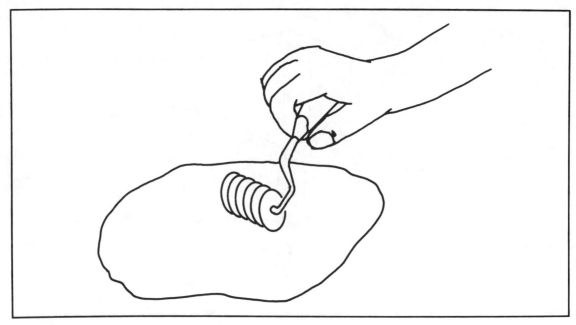

Fig. 16-12. Using a laminating roller.

material in large enough sizes. If joints are required, however, the basic technique is to overlap the pieces about 2 inches (Fig. 16-13). If mat is used, work down the laminate to a nearly even thickness with a dabbing action of the brush when applying resin to the overlap (Fig. 16-14).

If a second layer of reinforcing material is to be applied wet over the first layer (not required for the box project), position it in the wet resin and press it down and smooth it out. The brush can be used for this. If the second layer is cloth rather than mat, a brushing action can be used, but be careful not to unravel the cut edges of the cloth.

Saturate the second layer of reinforcing material with resin. Spread the resin as evenly as possible. Saturate all areas of the reinforcing material.

Apply all layers of reinforcing material that are to be applied wet at this time. When everything looks okay, clean the resin from the brush with acetone before the resin hardens on it.

It generally works best if laminating resin is used for all but the final layer of resin on the

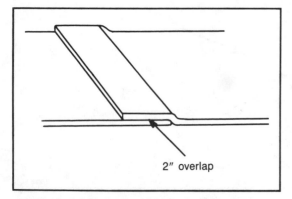

Fig. 16-13. Reinforcing material joined with overlap.

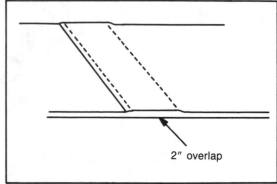

Fig. 16-14. Work mat overlap down to near even thickness.

230

backside of the laminate. Use sanding or finishing resin on the final layer to give a surface that is not sticky so that it can be sanded.

Allow the laminate to cure. Apply the next section of reinforcing material to the inside of one of the sides of the box in a similar manner. Position the box so that you will be fiberglassing downward on a level plane. Continue this process until the entire foam core is sheathed with a fiberglass skin.

If additional layer or layers of reinforcing material are to be added to the laminate (not required for the box project), measure out and catalyze the required amount of polyester resin. Brush on a layer of wet resin over the cured laminate. Position the first additional layer of fiberglass reinforcing material over the wet resin until it is centered, then press it down and smooth it out. The brush can be used for this.

Saturate the reinforcing material with resin. Spread the resin as evenly as possible. If additional layers are to be applied wet, these should be added at this time in a similar manner.

Allow the laminate to thoroughly cure. This completes the chemical part of the construction. To finish the box project, sand and fair the surfaces and then paint or apply color gel coat.

PLANTER BOX

A planter box (Fig. 16-15) is only slightly more difficult than the box project. Cut the bottom and four sides from a ⅜-inch thick sheet of rigid plastic foam to the dimensions shown in Figure 16-16. Epoxy glue the parts to form the box shape

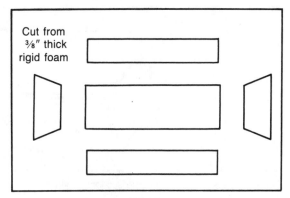

Fig. 16-16. Pattern for plastic foam parts for planter box.

(Fig. 16-17). Use a fiberglass putty fill to shape inside corners to a rounded contour (Fig. 16-18). Round outside angles with a file or coarse sandpaper. (Fig. 16-19). The edges of the lip of the box should also be rounded (Fig. 16-20).

To apply a fiberglass laminate, you will need laminating and finishing or general-purpose polyester resin, catalyst for the resin, mixing cups and sticks, a small brush, and sufficient 10-ounce fiberglass for sheathing the inside and outside of the planter box with two layers. Protect the work areas with heavy paper, wear proper protective clothing and safety equipment, and follow all safety rules.

Place the planter box form over the protective paper with the surface to laminate facing upward. For the planter box, the inside bottom surface is usually done first, followed by the inside sides, the lip, the outside sides, and finally the bottom. Allow the bottom inside surface to set up before doing

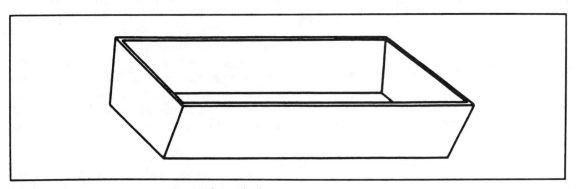

Fig. 16-15. Planter box layed up over rigid foam plastic core.

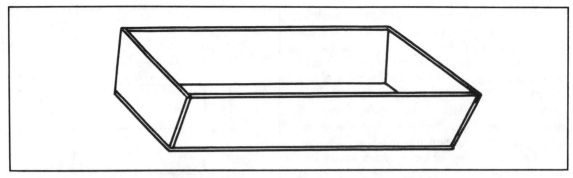

Fig. 16-17. Epoxy glue foam core sections together.

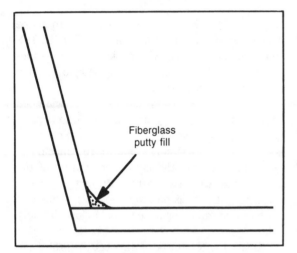

Fiberglass
putty fill

Fig. 16-18. Use fiberglass putty fill on inside corners to give a rounded contour.

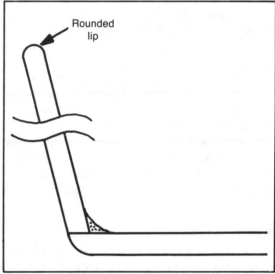

Rounded
lip

Fig. 16-20. Round the edges of the lip of the planter box.

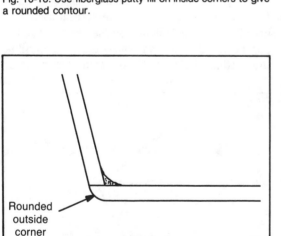

Rounded
outside
corner

Fig. 16-19. Round outside corners.

one of the inside sides, and so on. Because the fiberglass is to be bonded to the foam core, no mold release is applied.

To laminate the first section of fiberglass skin in place, cut the fiberglass cloth to size. For the inside bottom, this should be the size of the inside bottom plus an inch extra all the way around to extend upward on the sides (Fig. 16-21). Dry fit the cloth first. Make corner cuts in the fiberglass cloth.

The next step is to laminate the layer of fiberglass reinforcing material rigid foam plastic. Apply one layer to a section, allow this to cure, then apply

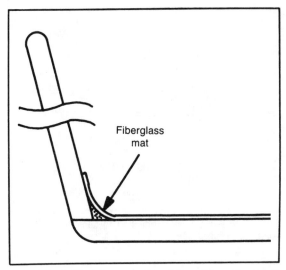

Fig. 16-21. Inside bottom section of fiberglass mat extends upward on sides an inch all the way around.

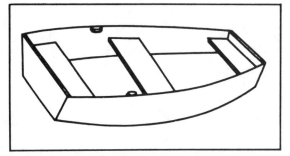

Fig. 16-22. Pram boat.

one layer to an adjacent section, and so on until the planter box is sheathed with one layer of fiberglass cloth. Then repeat this process for applying the second layer.

Next, bond the reinforcing material to the core material and saturate it with resin as you did for the fiberglass box. Allow the laminate to cure, then apply an adjacent section of reinforcing material to the inside of one of the sides of the planter box in a similar manner. Position the box so that you will be fiberglassing downward on a level plane. Continue this process until the entire foam core is sheathed with a fiberglass skin.

The second layer of the laminate is then applied in a similar manner. The final coat of resin should be finishing resin so that the surface can be sanded.

Allow the laminate to thoroughly cure. This completes the chemical part of the construction. To finish the box project, sand and fair the surfaces and then paint or apply color gel coat.

BOAT

A more difficult project is a small pram boat (Fig. 16-22). The bottom, sides, and bow and stern sections are cut from a ⅜-inch thick sheet of rigid plastic foam. The parts are then epoxy glued to form the boat shape (Fig. 16-23). The inside corners should be filled with a fiberglass putty fill to give a rounded contour, and outside angles can be rounded with a file or coarse sandpaper. The edges of the lip of the boat should also be rounded (see Figs. 16-18 to 16-20).

To apply a fiberglass laminate to the boat form, you will need laminating and finishing or general-purpose polyester resin, catalyst for the resin, mixing cups and sticks, a small brush, and sufficient 10-ounce fiberglass for sheathing the inside and outside of the foam core with four layers. Protect the work area with heavy paper, wear proper protective clothing and safety equipment, and follow all safety rules.

Place the boat form on the protective paper with the surface you want to laminate first facing upward. For the boat, do the inside bottom surface first, followed by the inside sides and stern and bow sections, the lip, the outside sides, and finally the bottom. Let the bottom inside surface set up before doing one of the inside sides and so on. Turn the form in a new direction for each lay-up so you

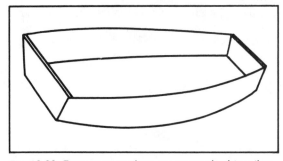

Fig. 16-23. Foam core sections are epoxy glued together.

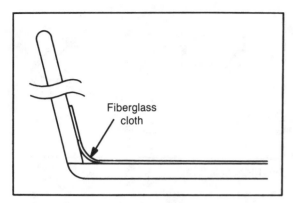

Fig. 16-24. Inside bottom section of fiberglass cloth extends upward on sides a couple of inches all the way around.

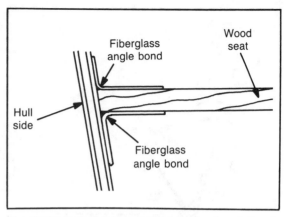

Fig. 16-25. Wood seat is attached to fiberglass hull with fiberglass bonding strips.

are working on a near level plane. Because the fiberglass is to be bonded to the foam core, no mold release is applied.

To laminate the first section of fiberglass skin in place, cut the fiberglass cloth to size. For the inside bottom of the boat, this should be the size of the inside bottom plus two inches extra all the way around to extend upward on the sides (Fig. 16-24). First dry fit the cloth. Make corner cuts in the fiberglass cloth.

The next step is to laminate the layer of fiberglass reinforcing material rigid foam plastic. For this project, it would works best to apply one layer to a section, allow this to cure, then apply one layer to an adjacent section, and so on until the core foam is sheathed with one layer of fiberglass cloth inside and out. Repeat this process for applying the second layer, then again for the third layer, and finally for the fourth layer.

It is usually best to use single pieces of fiberglass cloth whenever possible. It is also possible to join pieces, such as two pieces for the inside bot-

tom, with a lap joint (see Fig. 16-14). When the next layer is applied later, use a lap joint in a different area. The first piece of reinforcing material is their applied as detailed previously.

Allow the laminate to cure. Apply an adjacent section of reinforcing material to the inside of one of the sides of the boat in a similar manner. Position the boat so that you will be fiberglassing on as level a plane as possible. Continue this process until the entire foam core is sheathed with a fiberglass skin.

The second layer of the laminate is then applied in a similar manner. Follow with a third lay, and finally the fourth. The final coat of resin should be finishing resin so that the surface can be sanded.

Allow the laminate to thoroughly cure. To finish the boat project, sand and fair the surfaces. Wood seats can be attached to the fiberglass hull with fiberglass bonding straps (Fig. 16-25). The fiberglass hull can be finished by applying paint or gel coat, as desired.

Chapter 17

Construction Projects Using Contact Molding

A variety of interesting projects are possible using contact molding. Almost any shape is possible, but some are easier and more practical than others. As a rule, the molding becomes more difficult when the curves are greater and/or the mold is deeper.

CONTACT MOLDS

All of the projects detailed in this chapter require molding fiberglass by contact molding with one side against the mold surface. When one smooth surface is required, contact molding is used. Pressure molding, which gives smooth surfaces on both sides of the laminate, is more difficult and beyond the scope of this book.

Contact molds can be formed from a variety of materials. Fine-grained wood is frequently used. Because contact molding gives a smooth surface on one side only, it is important to note that there are two basic forms a deep rectangular contact mold can take. If you want to mold a box shape with a smooth inside surface, the contact mold surface will look like the outside of an upside down box (Fig. 17-1). If you want to mold a box shape with a smooth outside surface, the contact mold surface will look like the inside of a box with an open top (Fig. 17-2). To mold a box with a smooth inside surface, the mold is placed with the contact surface upward (see Fig. 17-1). The molding is then laid up over this (Fig. 17-3 and removed from the mold after it has cured (Fig. 17-4).

Contrast this to a mold for a box form with a smooth outside surface (see Fig. 17-2). The mold is positioned like an upright box with an open top and the molding is laid up inside (Fig. 17-5) and removed from the mold after it has been cured (Fig. 17-6). Depending how the molding is to be used, it can be positioned like a pan with a smooth underside and outsides or a cap or lid with a smooth top and outsides. In each case, it is important to keep in mind that the molding is not a duplicate of the mold but a reverse image.

Contact molds can be shaped from a single piece of wood or, more typically, from two or more pieces of wood glued together. Ordinary woodworking techniques are used for shaping deep rectangular molds from wood. Figure 17-7 shows a wood mold suitable for molding a box form with a smooth inside surface. Figure 17-8 shows a wood mold suitable for molding a box form with a smooth outside surface.

After the wood has been shaped and sanded, the next step is to prepare the wood surface so that it is suitable for molding. One method is to coat the molding surface with a thin layer of polyester resin. In addition to the shaped and sanded wood mold, you will need either finishing or general-purpose polyester resin, catalyst for the resin, a mixing container, short and long mixing sticks, and a small brush. Protect the work area with heavy paper, wear protective clothing and safety equipment, and follow all safety rules.

Position the wood mold with the molding surface upward. Open the resin container and stir

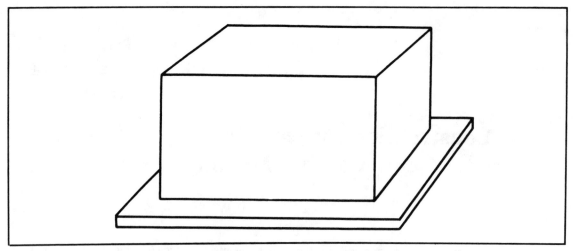

Fig. 17-1. Convex contact mold for box form gives molding with smooth concave or inside surface.

carefully with a clean mixing stick. Measure about 1 ounce of resin for each square foot of wood surface to be covered into the mixing cup. Replace the lid on the resin container and set it aside.

If available, add about ⅛ ounce of styrene monomer to the resin and mix thoroughly. While the resin can be applied to the wood surface without this, the thinning action of the styrene will thin the resin and allow it to better penetrate the surface of the wood.

Next, add 8 drops of catalyst for each ounce of resin to the resin. This is approximately 2 percent

by volume, which should give about 15 minutes of working time. Wait about 30 seconds, then brush a smooth thin layer of the catalyzed resin onto the molding surface (Fig. 17-9). When you have finished applying the coat of resin to the molding surface of the wood and everything looks okay, clean the brush with acetone. Allow the resin to harden on the wood surface. Even though this should happen about 15 minutes after the resin is applied, it is best to wait at least an hour before attempting to sand the surface.

Next, lightly sand the surface with fine-grit paper to remove the surfacing agent. This is not necessary if laminating resin is used. An alternate method is to remove the surfacing agent by wiping the surface with acetone on a rag. Then apply a second coat of polyester resin. When this has cured, apply a third coat.

When the third coat has cured, sand with fine grit wet/dry sandpaper and water. Polish the surface using a buffing compound by hand with a clean cloth or with a power buffer.

It is important to take the necessary time to make a good mold. Any defects in the mold surface will be copied on the moldings made with the mold.

A variety of existing forms can be used as molds. For example, pans made of stainless steel and aluminum are ideal. Many plastic pans will also work but make certain that the plastic is a type

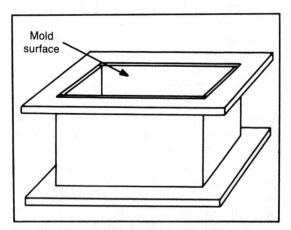

Mold surface

Fig. 17-2. Concave contact mold gives box molding with smooth convex side or outside surface.

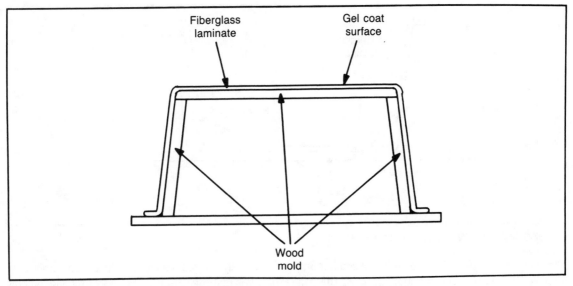

Fig. 17-3. Molding is laid up over convex mold form.

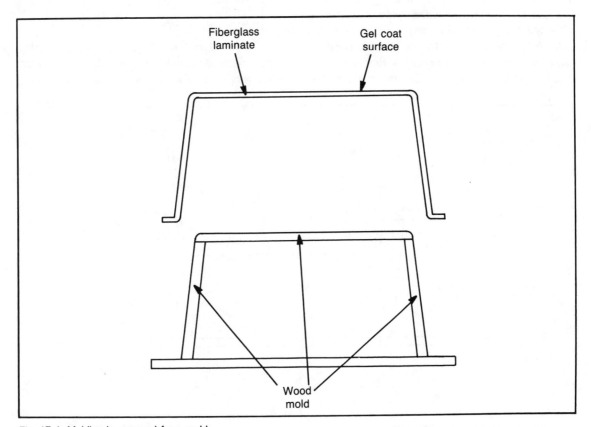

Fig. 17-4. Molding is removed from mold.

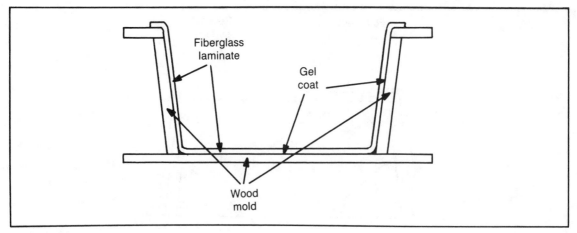

Fig. 17-5. Molding is laid up over concave mold form.

that polyester resin will not dissolve. Keep in mind that you will not be making a duplicate of the pan but using it as a mold to obtain a reverse image. Because you will be contact molding, you will be using one side of the pan when laying up a molding, although many pans will allow molding on either side. Make certain that the mold form does not have vertical sides or overhangs that will make removing a finished molding difficult or impossible.

Contact molds can also be made from a variety of other materials. For contact molding, you will be concerned with only one side of the mold, either for a concave or convex molding form.

CONTACT MOLDING

To do contact molding, you need a suitable mold, a paste-type mold release, polyester gel coat resin, laminating and/or finishing or general-pur-

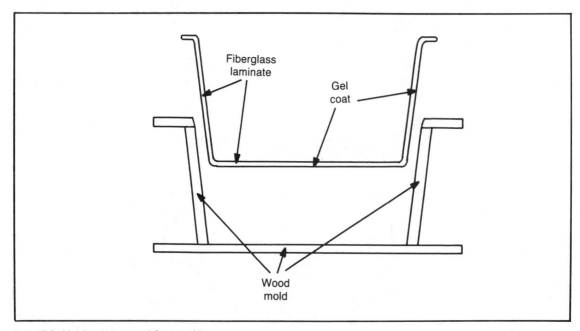

Fig. 17-6. Molding is removed from mold.

238

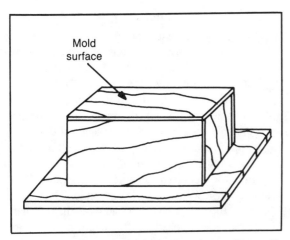

Fig. 17-7. Wood mold for laminating box shape with smooth inside or concave side.

pose polyester resin, catalyst, mixing cups, short and long mixing sticks, a small brush, and fiberglass reinforcing material (mat alone or in combination with cloth and/or woven roving is usually used for typical molding projects). Protect your work surface with heavy paper, wear protective clothing and safety equipment, and follow all safety rules.

Place the mold on the protective paper with the mold surface upward. If the molding surface of the mold is concave, do the lay-up inside a pan form. If the molding surface of the mold is convex, do the lay-up over a cap-like form. In either case,

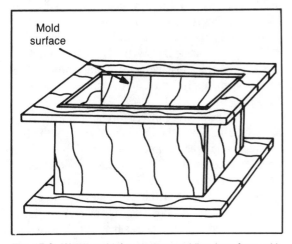

Fig. 17-8. Wood mold for contact molding box form with smooth outside or convex surface.

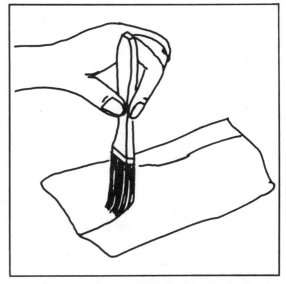

Fig. 17-9. Brush a smooth thin layer of catalyzed polyester resin on the wood molding surface.

apply a coat of the paste-type mold release, using a clean cloth to spread a thin layer of the release over the entire mold surface. The purpose of this is to keep the molding from sticking so that it can be removed after it has cured.

The next step is to apply the gel coat on the mold surfaces over the mold release. When you have everything ready, open the resin container. The resin can be clear or any desired color or have color pigments formulated for polyester gel coat resin added. Carefully stir the resin with a clean stick. Pour the required amount (this depends on area to be covered; usually about 1 ounce per square foot) of the gel coat resin into a mixing cup. Replace the lid on the gel coat resin container and set it aside.

Next, add the catalyst to the resin. This is approximately 2 percent by volume, which is usually used for polyester gel coat resin at 75 degrees Fahrenheit. This may vary, however, for the particular brand of gel coat resin being used. Follow the manufacturer's directions. Wait about 30 seconds, then brush a smooth thin layer of the catalyzed gel coat resin onto the mold surface over the mold release agent. Use a continuous motion with the brush (Fig. 17-10) rather than painting back

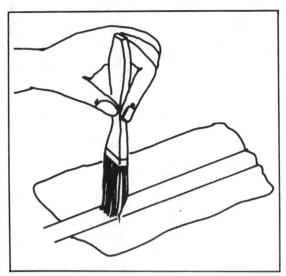

Fig. 17-10. Apply gel coat to the mold surface with long continuous strokes of brush.

and forth. The gel coat should have a thickness of only .02 to .03 inch. A thicker layer will be brittle and subject to cracking and crazing. When you have finished applying the coat of resin and everything looks okay, clean the paint brush with acetone. Allow the resin to harden.

Because gel coat resin usually doesn't contain a wax additive so that the laminate can be laid up directly over this with no additional surface preparation. Laminate the first layer of fiberglass reinforcing material (which is usually mat) to the gel coat. Cut the reinforcing material to a size slightly larger than is required to cover the molding area of the mold. Cuts with lap joints are usually required to form the dry mat to the approximate shape of the mold form. Beginners are advised to apply one layer of reinforcing material at a time and allow this to set before applying the next layer. As experience is gained, two or more layers (up to a maximum of about 6 ounces of reinforcing material per square foot) can be applied wet. This is usually more difficult than applying a single layer at a time but makes the work faster.

Mat can be formed to shallow curves without making cuts, but only after the resin has been applied to the material. Sharp curves and angles usually require making cuts and overlapping the mat in these areas. The overlaps can be worked

down to an even thickness after resin has been applied by dabbing with the brush.

Next calculate the amount of resin required for saturating the reinforcing material that is to be applied wet at this time (see Chapter 7). Pour the required amount of resin into a mixing cup. You will probably need about 7½ minutes of working time to apply the resin to a single layer of reinforcing material; more time may be required for multiple layers (again, see Chapter 7).

Add the required amount of catalyst to the resin and stir. Wait about 30 seconds. Then, use the brush to apply a coat of resin over the gel coat layer that was applied to the mold surface and allowed to set previously.

Position the first layer of fiberglass reinforcing material (usually a mat layer) over the wet resin until it is centered; press it down and smooth it out. For mat, use a dabbing action rather than a brushing action to avoid lumping up the mat.

Saturate the reinforcing material with resin (Fig. 17-11). Spread the resin as evenly as possible to saturate all areas, including areas that extend slightly beyond the desired size of the finished form. These will be trimmed off later after the laminate has cured.

For larger mold forms (which should not be attempted until considerable experience has been gained making small moldings), use a squeegee or laminating roller to remove air bubbles. Some molding projects require joints in the individual layers of reinforcing material. The basic technique is to overlap the pieces about 1 to 2 inches (Fig. 17-12). If mat is used, work the laminate down to a nearly even thickness with a dabbing action of the brush when applying resin to the overlap (Fig. 17-13).

If a second layer of reinforcing material is to be applied wet over the first layer, position it in the wet resin and press it down and smooth it out. The brush can be used for this. If the second layer is cloth rather than mat, use a brushing action but be careful not to unravel the cut edges of the cloth.

Saturate the second layer of reinforcing material with resin. Spread the resin as evenly as possible. Saturate all areas of the reinforcing material.

Apply all layers of reinforcing material that

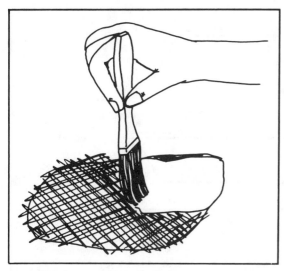

Fig. 17-11. Saturate the reinforcing material with resin.

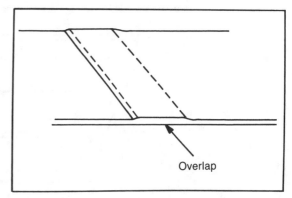

Fig. 17-13. Work mat overlap down to near even thickness.

are to be applied wet at this time. When everything looks okay, clean the resin from the brush with acetone.

It's best to use laminating resin for all but the final layer of resin on the backside of the laminate. Use sanding or finishing resin on the final layer to give a surface that is not sticky so that it can be sanded.

Allow the laminate to cure. If additional layer or layers of reinforcing material are to be added to the laminate, measure out and catalyze the required amount of polyester resin. Brush on a layer of wet resin over the cured laminate. Position the first additional layer of fiberglass reinforcing mate-

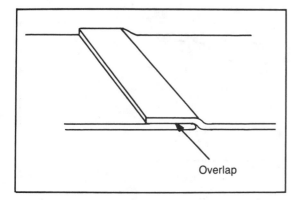

Fig. 17-12. Reinforcing material joined with overlap.

rial over the wet resin until it is centered; press it down and smooth it out using the brush. Saturate the reinforcing material with resin. If additional layers are to be applied wet, add them at this time.

Allow the laminate to thoroughly cure, then remove it from the mold. Even though the mold release was used, you might have to pry the laminate loose with a putty knife or other similar tool.

This completes the chemical part of the construction. For most projects, the fiberglass molding requires trimming. First mark pattern and saw off excess fiberglass at the lip of the pan form or elsewhere, as required. Sawing methods are detailed in Chapter 6.

File and sand the edges smooth. Polish and buff the gel coat surfaces and (for some projects) edges of the laminate.

DEEP RECTANGULAR PLANTER BOX

A deep rectangular planter (Fig. 17-14) is an interesting project. The planter box can be contact molded to give a smooth surface on one side only; either the inside or outside surface of the planter can be the smooth side, as desired.

The mold can be shaped from wood. Ordinary woodworking techniques are used for shaping and assembling the wood parts. Figure 17-15 shows a wood mold suitable for molding the planter box with a smooth inside surface. Figure 17-16 shows a wood mold suitable for molding the planter box form with a smooth outside surface.

After the wood has been shaped and sanded,

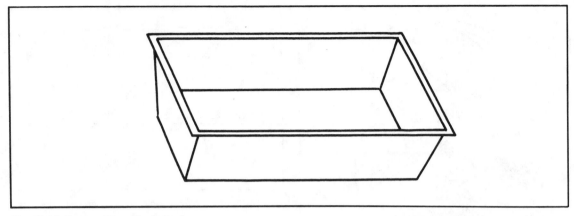

Fig. 17-14. Deep rectangular planter box.

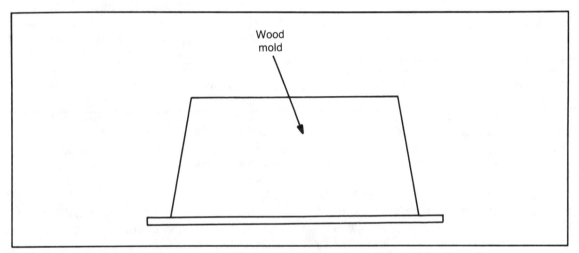

Wood mold

Fig. 17-15. Pattern for wood mold for contact molding rectangular planter box with smooth interior surface.

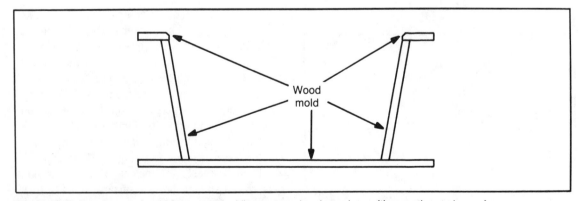

Wood mold

Fig. 17-16. Pattern for wood mold for contact molding rectangular planter box with smooth exterior surface.

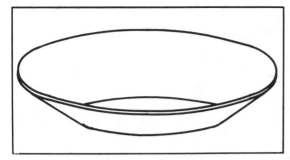

Fig. 17-17. Bowl.

Fig. 17-18. Wastebasket.

prepare the wood surface so that it is suitable for molding. One method is to coat the molding surface with a thin layer of polyester resin, as detailed previously in this chapter. After the polyester resin has cured, polish the surface using a buffing compound. This can be done by hand with a clean cloth or with a power buffer.

It is important to take the necessary time to make a good mold. Any defects in the mold surface will be copied on the moldings made with the mold.

To contact mold a planter box, you need a suitable mold, a paste-type mold release, polyester gel coat resin, laminating and/or finishing or general-purpose polyester resin, catalyst, mixing cups, short and long mixing sticks, a small brush, and fiberglass mat reinforcing material.

Apply the gel coat on the mold surfaces, using the same methods as detailed for the box. Allow the gel coat resin to cure. Gel coat resin usually doesn't contain a wax additive so that the laminate can be laid up directly over this with no additional surface preparation.

Next laminate the first layer of fiberglass mat reinforcing material to the gel coat, using the techniques given previously. The reinforcing material should be cut to a size slightly larger than is required to cover the molding area of the mold. Cuts with lap joints are usually required to form the dry mat to the approximate shape of the mold form.

Mat can usually be formed to shallow curves without making cuts but only after the resin has been applied to the material. Sharp curves and angles require making cuts and overlapping the mat in these areas. The overlaps can be worked

down to an even thickness after resin has been applied by dabbing with the brush.

Saturate the reinforcing material with resin. Spread the resin as evenly as possible to saturate

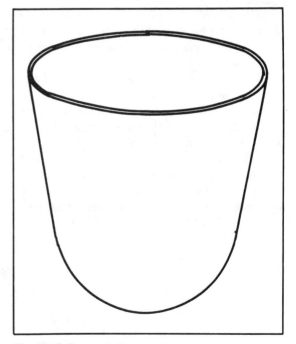

Fig. 17-19. Deep cylinder.

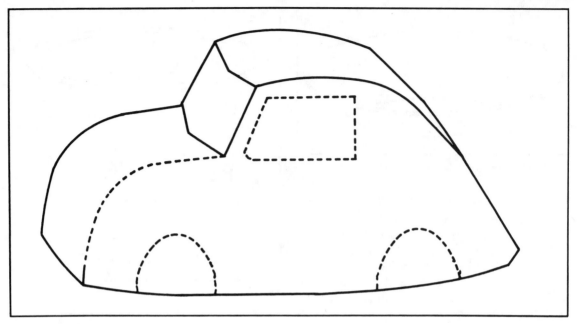

Fig. 17-20. Toy car body.

all areas evenly, including parts that extend beyond the finished size. These will be trimmed off later after the laminate has cured.

Use laminating resin for all but the final layer of resin on the backside of the laminate. Use sanding or finishing resin on the final layer to give a surface that is not sticky so that it can be sanded.

Allow the laminate to cure. Then apply a second layer of mat over the first layer. Allow the laminate to thoroughly cure, then apply a third layer to the laminate in the same manner. The final coat of resin should be finishing resin.

Allow the laminate to thoroughly cure, then remove it from the mold. Even though the mold release was used, you might have to pry the laminate loose with a putty knife or other similar tool.

This completes the chemical part of the construction. To finish the planter box, mark pattern and saw off excess fiberglass at lip, as required. Sawing methods are detailed in Chapter 6. Next, file and sand the edges smooth. Polish and buff the gel coat surfaces and edges of the laminate.

OTHER MOLDING PROJECTS

A variety of other forms can be contact molded in a similar manner, including a bowl (Fig. 17-17), a wastebasket (Fig. 17-18), a deep cylinder (Fig. 17-19), and a toy car body (Fig. 17-20). Begin with small projects with simple shapes and then gradually work up to larger and/or more difficult ones.

Appendix

Appendix: Suppliers

Here is a listing of suppliers of all types of fiberglassing materials and kits.

MAIL-ORDER FIBERGLASSING MATERIALS AND SUPPLIES

ClarkCraft,
16-R1 Aqualane,
Tonawanda, NY 14150

Polyester and epoxy resins; fiberglass cloth, mat, and woven roving; plastic foams; etc.

Defender Industries, Inc.
255 Main St.
New Rochelle, NY 10801

Offers complete line of fiberglassing materials and supplies at discount prices; also has large selection of marine equipment, hardware, and supplies. Their 168-page catalog is $1.

RETAIL FIBERGLASSING MATERIAL AND SUPPLY STORES

TAP Plastics, Inc., offers a complete line of fiberglassing materials and supplies at the following retail stores:

1401 N. Clovis Ave., #101
Fresno, CA 93727

1212 The Alameda
San Jose, CA 95126

4227 Pacific Ave.
Stockton, CA 95207

3011 Alvarado St.
San Leandro, CA 94577

2041 East St.
Concord, CA 94520

4538 Auburn Blvd.
Sacramento, CA 95841

606 South B St.
San Mateo, CA 94401

12404 N.E. Halsey
Portland, OR 97230

2945 S.W. Temple
Salt Lake City, UT 84115

Glen-L Marine
9152 Rosecrans
Bellflower, CA 90706

Complete line of fiberglassing materials and supplies. Also available by mail-order.

OTHER FIBERGLASSING MATERIAL AND SUPPLY SOURCES

Aladdin Products, Inc.
RFD 2
Wiscasset, ME 04578

STR-R-ETCH MESH and FER-A-LITE.

American Klegecell Corporation
204 North Dooley St.
Grapevine, TX 76051

Klege-Cell polyvinyl chloride rigid foam core material in four different densities ranging from 2.0 to 15.0 lb./cu. ft.

Baltek Corporation
10 Fairway Center
Northvale, NJ 07647
Balsa core material.

Devcon Corporation
Endicott St.
Danvers, MA 01923
Underwater epoxy.

Dynatron/Bondo Corporation
2160 Hills Ave., N.W.
Atlanta, GA 30318
Auto body plastic fillers and fiberglassing repair materials and kits.

Fibre Glass Evercoat Company
6600 Cornell Rd.
Cincinnati, OH 45242
Resin, fiberglass reinforcing materials, filler materials, and repair kits.

Lan-O-Sheen
1 W. Water St.
St. Paul, MN 55107
Resin and fiberglass reinforcing materials.

Lonza, Inc.
22-10 Rt. 208
Fairlawn, NJ 07410
Airex polyvinyl chloride rigid foam core material.

Magic American Chemical Corporation
23700 Mercantile Rd.
Cleveland, OH 44122
Fiberglass cleaner.

Oatey Company
4700 W. 160th St.
Cleveland, OH 44135
Bond Tite auto body plastic fillers and fiberglassing repair materials and kits.

Pettit Paint Company, Inc.
36 Pine St.
Rockaway, NJ 07866
Complete line of marine paint, including two-part polyurethane, resins, fiberglass reinforcing materials, and fiberglass repair kits.

Plastic Sales and Manufacturing Company, Inc.
3030 McGee Trafficway
Kansas City, MO 64108
Polyester and epoxy resins and fiberglass reinforcing materials.

Ram Chemicals
210 E. Alondra Rd.
Gardena, CA 90248
Gel coat resins and release agents.

Rule Industries Inc.
Cape Ann Industrial Park
Gloucester, MA 01930
Resins, fiberglass reinforcing materials, and fiberglass repair kits.

Travaco Laboratories, Inc.
345 Eastern Ave.
Chelsea, MA 02150
Marine-Tex epoxy repair compound.

Woodhill Chemical Sales Corporation
18731 Cranwood Parkway
Cleveland, OH 44128
Duro Black Knight auto body plastic fillers and fiberglassing repair materials and kits.

FIBERGLASS AUTO AND VAN COMPONENTS

A & A Fiberglass, Inc.
1534 Nabell Ave.
Atlanta, GA 30344
Auto repair and customizing components.

Fiberfab, Inc.
548 Baldwin St.
Bridgeville, PA 15107
Auto repair and customizing components.

Mansfield Fiberglass & Plastic Company
P.O. Box 127
Thorntown, IN 46071
  Auto repair and customizing components.

Douglass Super Top
P.O. Box 701
Valley Forge, PA 19482
  Fiberglass van tops for converting vans to standup headroom campers.

Motion Performance Parts, Inc.
598 Sunrise Highway
Baldwin, L.I. NY 11510
  Auto repair and customizing components.

Perfect Plastics Industries, Inc.
Schreiber Industrial District, Building 213
New Kensington, PA 15068
  Auto repair and customizing components.

FIBERGLASS BOAT KITS

Aquaglas Industries
67 Kennedy Ave.
Blue Point, NY 11715
  16-foot runabout.

Dreadnought Boatworks
P.O. Box 221
Carpinteria, CA 93013
  32-foot Dreadnought world cruising type sailboat.

Glander Boats, Inc.
P.O. Box 1107
Tavernier, FL 33070
  23-foot and 33-foot sailboats.

Heritage Marine, Inc.
2919 Gardena
Long Beach, CA 90806
  27-foot Nor'Sea cruising sailboat.

Lakeview Boat Company
P.O. Box 5595
Riverside, CA 92507
  Canoes and Kayaks.

Luger Industries, Inc.
3800 West Hwy. 13
Burnsville, MN 55337
  Sailboats ranging from 16 to 30 feet in length and powerboats ranging from 21 to 30 feet in length.

Pacific Seacraft
3301 South Susan St.
Santa Ana, CA 92704
  Flicka 20, Seacraft 25, and Mariah 31 cruising sailboats.

Roberts & Matthews Yacht Corp., Inc.
P.O. Box 10324
Brandenton, FL 33507
  Adventure 25, Roberts 26, and Roberts Spray 40 sailboats.

Trailcraft, Inc.
P.O. Box 606
Concordia, KA 66901
  Kayaks, canoes, and crossbreeds.

Yacht Constructors, Inc.
7030 N.E. 42nd Ave.
Portland, OR 97218
  Cascade 27, 29, 36, 42, and 43 sailboats.

Yachtcraft Corp.
551 W. Crowther Ave.
Placentia, CA 92670
  Yachtcraft 30, 34, 37, 40, 41, and 44 sailboats.

FIBERGLASS KIT CARS

Anderson Industries, Inc.
5625 Furnace Ave.
Elkridge, MD 21227
  Roadster car bodies and kits.

Elite Enterprises, Inc.
690 E. Third St.
Cokato, MN 55321
  Sports car bodies and kits.

Perfect Plastics Industries, Inc.
Schreiber Industrial District, Building 213
New Kensington, PA 15068
    Car bodies and kits for VW chassis.

FIBERGLASS TRAVEL TRAILER KITS
Burro Company
14143 21st Ave. North
Plymouth, MN 55441

Glossary

Glossary

accelerator—A highly active oxidizing material such as cobalt that is added to polyester resin to produce internal heat, so the resin will cure at room temperature.

acetone—A cleaning solvent for removing uncured resin from brushes and tools.

air-inhibited resin—A resin in which the presence of air will inhibit the cure of the surface. The surface usually becomes hard but is tacky.

ambient temperature—Surrounding temperature or "room" temperature.

barrier cream—A skin cream used to protect the skin from possible contact with resins.

binder—An adhesive that is soluble in resin and is used to loosely bind glass fibers together to form fiberglass mat.

catalyst—Component added to polyester resin to initiate the curing, usually by oxidizing an accelerator.

cavity—A female mold or the laminating space between matched molds.

chopped strands—Glass fiber strands chopped up into short lengths.

close weave—Reinforcing fabric with the woven strands almost touching.

color pigments—Pigments that are added to resin to change its color.

core—Material used between two fiberglass skins to space them apart and give greater stiffness.

crazing—Tiny cracks in the surface of a fiberglass molding.

cure—The process of resin changing from a liquid to a solid state.

cure time—The time from when the catalyst or hardener is added to a resin until the resin reaches a cured state.

epoxy resin—A resin that is stronger and has better physical properties than polyester resin but is more difficult to use and considerably more expensive.

exothermic heat—Developed within the resin.

feathered edge—Tapered edge of fiberglass laminate.

fiberglass—Fine fibers of glass; reinforcing materials made from glass fibers; laminates of glass fiber reinforcing material and cured resin.

filler—Substance added to resin to form a putty; a resin filler or putty.

finish—Applied to glass fibers to allow resin to flow around and adhere to the fibers.

foam—A rigid plastic material that is very light in weight.

foam core—Foam used as a core material between two skins of fiberglass.

gel—A semisolid or jellylike state of resin when partially cured.

gel coat—Surface coat of resin that does not contain glass fibers and is usually colored.

glass fiber—A fine fiber of glass.

hardener—Component added to epoxy resin to initiate the curing.

lamination—Layers of glass reinforcing materials and resin that form a fiberglass panel.

lay-up—Process of applying resin to reinforcing materials placed in a mold.

253

mold—Form used for fiberglass lay-up to give desired shape and surface.

mold release—Substance used to prevent molding from sticking in the mold.

molding—Cured fiberglass object that has been formed in a mold.

nonair-inhibited resin—A resin that gives a surface cure in the presence of air by excluding air from the surface of the resin.

open weave—Reinforcing fabric with considerable space between woven strands.

plain weave—Common over the under weave used for making fabrics.

plastics—Synthetic materials; sometimes used to mean "fiberglass."

polyester resin—Resin commonly combined with reinforcing materials to form fiberglass. Because of lower cost, it is usually used instead of more expensive epoxy resin.

pot life—The length of time that a resin remains usable in a container after catalyst or hardening agent has been added.

putty—A resin filler material.

release agent—A coating applied to mold to prevent molding from sticking to the mold.

resin—A liquid plastic substance that cures to a hard substance when a catalyst or hardener is added; combined with reinforcing materials to form fiberglass.

roving—Continuous strands of glass fibers used to form untwisted yarn, which can be woven into woven roving reinforcing material.

sandwich construction—A core material with fiberglass skins.

shelf life—The length of time uncatalyzed resin will remain usable when stored in a sealed container; also applies to paints and other substances.

styrene—Liquid plastic used to thin polyester resin.

surfacing agent—Oil or wax material that goes to the surface of polyester resin during cure to inhibit air.

tack-free—A surface that is not sticky.

tacky—Sticky.

thixotropic—A liquid that has a high viscosity, so it will not flow easily.

thixotropic paste or powder—Added to resin to increase viscosity.

undercut—Reverse draft in a mold.

unidirectional—Strength is mainly in one direction; often applied to a reinforcing material that is woven to give greater strength in one direction than another.

vacuum bag molding—A method of molding that uses a flexible bag and a vacuum.

viscosity—Degree to which a liquid resists flow.

warp—Fibers woven across a fabric.

woven roving—Reinforcing fabric woven from strands of rovings, which are untwisted groups of glass fibers.

yarn—Twisted strands of glass fibers that are woven to form cloth.

Index

Index

Other Bestsellers of Related Interest

The Fiberglass Boat Repair Manual
Allan H. Vaitses
This is the definitive book on the subject, covering not just cosmetic dints and scratches, but also major repairs of structural damage to hulls and decks, delamination, refinishing, blistering, and more. A manual for the owner of a sail or powerboat, large or small, who desires an encyclopedic source for the care of his or her boat. Also invaluable for anyone seeking to buy and restore an old and perhaps damaged boat.
0-07-156914-6 $29.95 Hard

Sailboat Hull & Deck Repair
Don Casey
Everyone knows that fiberglass is durable, malleable, and easy-to-maintain, but what is not generally known is that it is also easy to repair. Among the repairs covered are rebedding deck hardware, replacing portlights, fixing leaky hull-deck joints, and repairing holes, cracks, blisters, and gouges. With the help of this manual, you will find that there is almost no fiberglass boat that you can't do as well as a boatyard.
0-07-013369-7 $19.95 Hard

Fiberglass Boatbuilding for Amateurs
Ken Hankinson
The only title available on home building of fiberglass boats. The author assumes you have no previous experience with fiberglass building and so covers the subject comprehensively. Detailed information on tools, equipment, materials, resin, fillers, plans patterns, finishing, foams, sandwich cored hulls . . . right through to outfitting. Includes a glossary, bibliography, and a listing of suppliers and manufacturers.
0-07-158235-5 $29.95 Hard

The Pool Maintenance Manual
Terry Tamminen
All aspects of pools, spas, and other recreational water containment units are covered in detail, from design and construction to lighting, winterizing, fiberglassing, tiling, and troubleshooting and repair. Major topics include basic and advanced plumbing systems; motors, pumps, horsepower, and hydraulics; filters and waterheaters; and water chemistry. This book is also an invaluable reference for the pool or spa owner.
0-07-061408-3 $24.95 Paper

Builder's Guide to New Materials & Techniques
Leon A. Frechette
This new guide focuses on innovative building materials, describing the best ways for contractors and subs to use these products to outperform competitors. From environment-friendly engineered lumber to energy-saving, concrete/masonry homebuilding systems, this unique book addresses the latest technologies and provides field-tested advice on tools and techniques. The book offers ideas for alternatives to traditional materials used in virtually every job.
0-07-015763-4 $44.00 Hard

Drywall Construction Handbook

Robert Scharff

This book is a must for residential and commercial drywall contractors. It covers the latest state-of-the-art materials that meet or exceed new fire regulation codes. Readers will find details on installing wood and metal-stud drywall partitions, techniques for applying drywall boards to masonry walls, and the various types of compounds used to tape joints. Even the most experienced drywall contractors will find useful tips and techniques in this informative handbook.

0-07-057124-4 $49.95 Hard

Roofing Handbook

Robert Scharff

Scharff covers everything roofers need to know to stay ahead of their competitors and avoid costly mistakes. He describes new, money-saving techniques and alternative materials, and offers valuable advice on issues such as bidding effectively, choosing suppliers, buying materials in bulk, and budgeting. Chapter topics include the major types of roof coverings and various roof drainage and ventilation systems.

0-07-057123-6 $49.95 Hard